BEE-MASTER REVISITED

The Autobiography of

GEORGE WAKEFORD

with

further biographical commentary by

Geoffrey Lawes

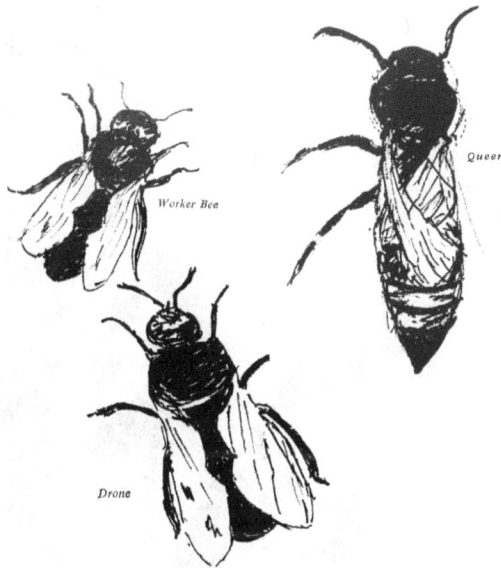

Illustration is of the three castes of bee as drawn by Agatha Bowley.

Northern Bee Books

BEE-MASTER REVISITED

The Autobiography of

GEORGE WAKEFORD

with

further biographical commentary by

Geoffrey Lawes

Northern Bee Books

BEE-MASTER REVISITED

The Autobiography of GEORGE WAKEFORD

© Geoffrey Lawes

ISBN 978-1-908904-12-6

Published by Northern Bee Books, 2012
Scout Bottom Farm
Mytholmroyd
Hebden Bridge
HX7 5JS (UK)

Design and artwork
D&P Design and Print
Worcestershire

Printed by Lightning Source, UK

Contents

Introduction vii

Foreword ix

Chapter I *Boyhood* 1

Chapter II *Life at Ebernoe* 19
Thrashing
Tree Felling

Chapter III *My Beginning with Bees* 55
The Lore of Bees
My Garden

Chapter IV *Tales of West Sussex* 57

Appendices *I Article for the* Daily Mail, ... 109
II Poems,
III Flowers Favoured by Bees.
IV

Acknowledgements 119

Introduction

Signpost to Ebernoe at Balls Cross

Mr. George Wakeford's short autobiography, originally written in cooperation with Agatha H. Bowley and published at his own expense in 1977, was issued twice in limited numbers and has long been inaccessible to new readers.

In reviewing the work with a view to a third issue, the publishers deemed it respectful to his memory to include details of his life as a Bee-master which the man himself had been too modest to include in his own story. George's sentences, plain and characterful as they are, fall short of doing him full justice as a fine country gentleman and respected master of beekeeping, an omission we hope to have remedied.

Realising that George's own words, expressed in his unsophisticated Sussex voice, have a unique charm that must be recognized, identified and valued for their own sake, we determined that they should appear distinct from additional matter supplied by the revisers. Accordingly they are here printed in a **bold Times Roman font**. The new material is readily distinguishable by its more formal style, a smaller typeface and the use of the third person singular, rather than the first person.

The original photographs and Agatha H. Bowley's drawings are reproduced in this revised edition, together with many fresh photographs and illustrations intended to enhance the text. In so doing an attempt has been made to explain some of the beekeeping developments which George witnessed and encouraged in his lifetime which may add interest to modern beekeepers who were denied the privilege of knowing Mr. Wakeford. Additional information, intended for readers not well schooled in beekeeping matters, is included where it may add to general interest. The common reader will find this account of the life of an eccentric country character to be of absorbing interest and the history and geography of George's haunts adds another dimension to the appeal of this unusual book.

FOREWORD

Mr. George Wakeford, or George as he is better known to his many beekeeping friends, has given me the honour of introducing this book.

George himself needs no introduction as he is well known over a large part of the country for his proficiency in handling his friends, the bees.

The "Bee Master" – could there be a more appropriate title for this grand old Sussex gentleman?

When first asked I anticipated a book on beekeeping, knowing how much valuable knowledge he could have written for posterity. However he has chosen to write first what might be termed his Life's History, and told it in his most typical and delightful Sussex dialect, making it most fascinating.

His beekeeping methods are well known, but this story from boyhood onwards will no doubt give an insight into those days and times not so well known to many, and should give pleasure and interest to those who read this book.

ARTHUR S. CURNUCK,
Lecturer in Beekeeping
West Sussex County Council

[Mr. Curnuck was the last County Advisor appointed by the Council to advise schools about beekeeping, when Rural Studies was an important part of Secondary Schools' curricula and Beekeeping courses were offered to adults in Evening Classes. The post and the practice have now been discontinued. George was himself an occasional lecturer at Evening Classes.]

Chapter I

BOYHOOD

PALFREY'S FARM AT KIRDFORD

MR. Harold Roberts, Wisborough Green, who sends this sketch, notes that Palfrey's Farm, Kirdford, which lies back from the main London road towards Balls Cross, is very ancient, and, as he indicates in one corner, that it contains an interesting fire-back which is one of the oldest in the district. It has a double front door with Tudor fittings.

A drawing of Palfry Farm by neighbour Harold Roberts. Inset shows ancient fireback

I was born at Palfrey Farm near Petworth, a farm of one hundred and thirteen acres, originally built in 1327. We didn't see very many people at that time because we were a mile off the main road. We couldn't see another house from where we lived. The most we saw was the postman, the grocer and the man who came to get chickens, eggs and rabbits.

A drawing of Palfrey by A. Bowley

George's date of birth was 7th January, only a week into the 20th century about a year before the death of Queen Victoria and the beginning of the prosperous Edwardian interval before the Great War which broke out when he was fourteen. He would often remark on his own good fortune in being just too young for World War 1 and just too old for anything but the Home Guard in 1941. Palfrey farm is in Ebernoe, a hamlet of the village of Kirdford with a population of 261 in 1901.

In the first months of George's life Ladysmith was relieved, heralding the end of the Boer War and Sir John Lubbock, first President of the British Beekeeping Association (BBKA), who gave us bank holidays, was ennobled. Lord Leconfield, aged 24, came into his title at Petworth House, famous for its landscaping by Capability Brown and its paintings by J.M.W. Turner. British power in the world was at its highest point, a dominance of Empire which was set to decline steadily throughout the 85 years of George's life. In his lifetime the heads of six monarchs, two Queens and four Kings, appeared on British postage stamps.

He was christened into the Church of England as Edward Henry George Wakeford, being given the forenames of English kings, but it was only in official registry documents that he was known as Edward. The name of his father was always his familiar Christian name. Palfrey Farm may take its title from the term for a small saddle horse, most often ridden by ladies in olden times, but more probably arose from an early 14th century occupant named Richard Palfreyman. The farmhouse is the oldest building in the village. On the evidence of tally marks on a bench it is supposed that the house was once a coaching inn. Some of the early owners were probably ironmasters when the happy conjunction of timber for making charcoal, a controllable water source needed to power the bellows for the smelting furnace and forge hammer and accessible iron ore seams, promoted the local iron industry from the time of Queen Elizabeth I. The Penycod family owned and farmed Palfrey from 1541 to about 1725. They were maltsters. One of them was reputed to be a smuggler, and became known about 1800 as a 'hawker', probably dealing in silk, brandy and Hollands gin!

The Peacheys were Lords of the Manor who owned the land farmed as tenants by the Wakefords. Edmund Peachey, who died in 1656, a Sussex man, made money in London as a haberdasher, as did his son Edward as a wholesale grocer, mainly from pepper and spice. He became the first Lord of Ebernoe. All these magnates made property deals in the Petworth area and contracted youthful judicious marriages to add to their wealth. By 1829 they owned 900 acres. Edward's grandson, the second William, had at least two 'wives' married under dubious circumstances in the Fleet prison. He also kept a mistress, Elizabeth Paine, daughter of a Northchapel blacksmith, to whom he left his estate. The next William, the natural son, built Ebernoe House at a time when it was fashionable for rich people to indulge themselves with a gracious building. His son, William Graccus Peachey, so named to secure a legacy, was declared 'of unsound mind' by a Chancery Commission at Petworth. His brother Rev John, who was made responsible for administration, became Rector of Alfold and Lord of the Manor in 1846. His son, the last William, revived the famous Horn Fair. It is recorded that old Daniel Wakeford elected to walk home across the fields to Butcherlands one stormy night after a tenants' dinner at Ebernoe House, lost his tall hat in the wind, chased it, stumbled into a ditch and was drowned. Through William and Lord Leconfield's good offices Ebernoe Church was built in 1865 and also the Vicarage. He was succeeded by his brother John.

Both took a keen squire-like, paternalistic interest in the village, though their personal fortunes steadily waned. John's fondness for stag hunting and his own pack of harriers endeared him to many but were an expense he could ill afford. By the time he took on the estate, with the role of 'Squire', it was a poisoned chalice. John had four daughters but all died childless. They all moved away in 1911 when John died and the estate was bought into Lord Leconfield's Estate [Charles Henry Wyndham, 3rd Baron Egremont 1872 -1952 Lord Lieutenant of Sussex 1917-1949]. Ebernoe House was then sold off as a private dwelling being undermined by rabbits, in poor repair and of no use to My Lord. The whole estate had become dilapidated by immutable outgoings totally unmatched by income and by lack of sound management.

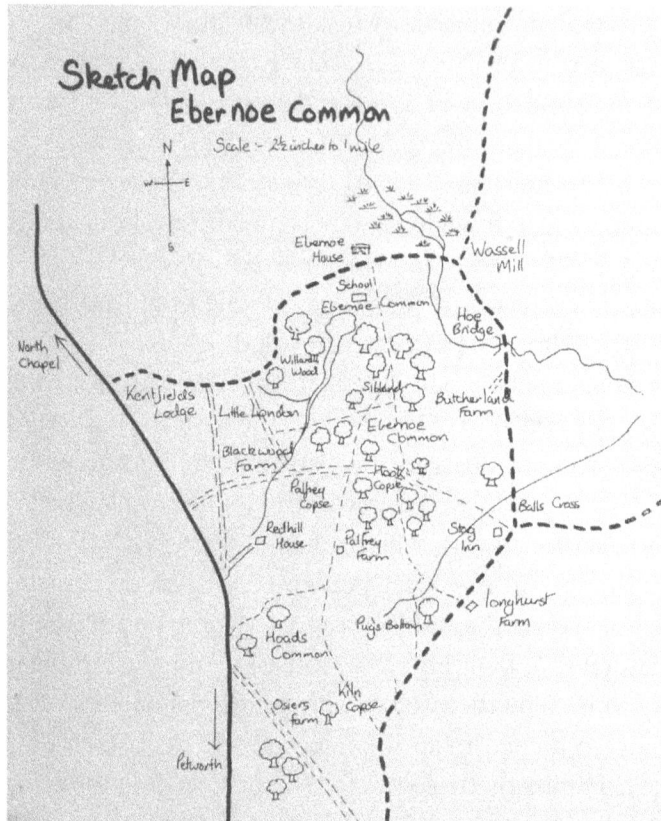

Map of the Ebernoe area by A. Bowley

Palfrey Farm was badly run down when George senior took it on. He worked all hours to survive and maintain his animal stock. In the very month of young George's birth he was obliged to write to the agents asking for extra time to pay his rent

– 'I have not got the money for my hay and my pens are full of porkers all not quite ready to go off'.

Farming and the profit of land had gone into serious and ever-deepening recession since 1874. The effects of free trade and cheap imports from the new world had crippled primary producers of corn and meat. Rural poverty forced thousands to emigrate and thwarted efforts of good husbandry. Land was cheap to rent, letting hard to achieve, especially for property with tumbledown buildings and difficult clay soils. Men like old George worked themselves into an early grave and the landowners gained insufficient rents to sustain their property or pay their bills. Palfrey Farm would have been held on the usual annual form of tenure, a tenancy at will, subject to six months' notice. Responsible landlords felt obliged to avoid disturbing tenants allowing continuity for sons and widows, especially when they accepted his politics, his interest in game and sport and conformed to the accepted local standards.

Here is an abbreviated version of more words provided by George to the Petworth Society Magazine in 1983.

"My father was working as a poultry boy for Lord Leconfield but he always wanted to farm Palfrey…he had to save up…by working evenings in the woods. He would take a 2lb jam jar, put a candle in it…and make hoops for barrels and 'hoop chips' –cleft pieces of wood, a by-product of the hoops themselves and were used for tacking round boxes to strengthen them…He would be paid so much for the bundle. He worked on regardless of the weather and careful of the sharpness of the cutting tool.

[Henry Puttock had a wholesale depot, The Hoop Sheds, at Billingshurst Station to collect and distribute coppiced wood. Hoops were used round barrels in the fish and West Indian sugar trades.]

Hoop-making

A bundle of hoops

He needed the money to buy stock, not for the rent. Palfrey was one of Squire Peachey's ten farms. Previous tenants were the Pyecrofts, licensees of the Swan and Half Moon in Petworth, the farm being run by a manager. My father was told that he could go there but would probably starve if he did. The rent was quite low, £40 a year for 113 acres. It wasn't easy to get people to take farms on. The house…was very dilapidated and the fields empty. The only thing there was in plenty were rabbits. If my father was in danger of starving he could always go to the hedgerow and shoot a couple. He was not allowed to touch the pheasants. In spite of warnings he became inseparable friends with Mr. Turner the Peachey Estate gamekeeper…a strict and incorruptible man.

In these days before Palfrey went over to Leconfield a year or two before the Great War the cellar at the house was floored simply with bricks laid upon the bare soil…the earthworms would throw up their casts between the bricks. Lord Leconfield later had the floor cemented. If father wanted timber for reparations he would get permission from Mr. Turner, cut the timber and cart it to Palfrey himself. Mr. Brown, the Estate carpenter, would do the repair. Mr. Peachey was a tall man …He didn't however tour his farms. The staff at Ebernoe, besides Mr. Turner and Mr. Brown, was Edgar Feist, the gardener, Mrs. Jupp, the housekeeper and David Baker of Golden Knob, a younger man who worked in the gardens.

Squire Peachey's hare-hounds weren't actually beagles so much as small fox-hounds…not many…perhaps a dozen. I don't think the hare-hounds had any uniform, though they probably once did. If they had been disbanded it would have been a break with tradition, the end of an era. Feist, the whipper-in, had a long whip and Turner, the huntsman, a hunting horn. There were never many followers…the most regular was Bill (Sticky) Knight, who used to be an umpire at cricket and walked on two crutches. They didn't usually catch the hare. It was the pursuit that really mattered. I remember them once catching a hare at Palfrey. It was a mistake. They didn't give it a proper start. They paunched it and threw it in the air for the hounds to fight over. Father could earn more taking the hounds over to Fittleworth in a high-sided cart than working on the farm all day.

My father would never catch a hare but the foxes certainly would and he had a secret agreement with the squire to catch them. Lord Leconfield would have objected most violently had he known. The traps were double-sprung on each side and nasty things to set. He had to kneel on the spring to get the jaws to open. We'd then put out the meat, dig a shallow hole and set the traps round the hole. Foxes are very cagey creatures and we'd set a decoy with freshly forked ground and meat – they'd often avoid that and go for the trap. In the morning my father would either shoot them or kill them with a club. If the foxhounds were due we'd cover the traps with a hurdle, fox traps were certainly not something to make public. [A tenant caught killing game pheasant, hare, fox or stag was liable to forfeit his tenancy! Lord Leconfield was an enthusiast for fox and stag hunting].

Ebernoe House, rather unusually, had its own rabbit warren, right up to the house itself. Rabbits were usually to be seen, black, brown an all shades in between. Mr. Turner would net and cull them regularly. The warren was like an open meadow running from the rhododendrons to the road.

Squire John Peachey inspects his rabbit warren at Ebernoe House

Father had two horses to begin with…but much of the work was done by hand. My mother brought her goats from Lodsworth but they had to be got rid of… They had eaten the bark off the holly trees so they died. We'd reap by hand with sickle or scythe. For peas and tares we'd prefer the scythe…they'd be roughly bundled then stacked in small ricks and fed to the animals while still green. After that we'd feed rye and lucerne. My mother would make 30lbs of butter a week, all packed neatly in half pound packs, readily sold to private customers or taken into Messrs. Gordon Knights at Petworth on a push-bike. Another 'hand-job' was flailing beans, done in the barn…exhausting work. We might also use the winnowing machine for removing the chaff from beans. The flails were fashioned locally. Sometimes our grass seed would be adulterated with a strange parasitic plant called 'dodder' which lives by attaching itself to a host plant and throwing out suckers…

In the very early days we'd take bullocks to Pulborough on foot. Once my brother and I were each given charge of a bullock to go the nine miles to market. Mine went faster than his. I got to Pulborough, had the bullock weighed and set off home on foot. I met my brother at Stopham Bridge with the bullock lying resting in the road. It was too fat and had had no exercise…Later we would borrow Lord Leconfield's bullock cart and two horses to pull it. Chicken, ducks, geese and eggs were taken to Brighton by Mr. Bridger, the Fittleworth carrier, who collected from all the local farms.

The water at Palfrey was very hard, but made good beer. We used to pick up the hops at Coultershaw Mill in big sacks, together with the malt. We'd pick it up when we collected the flour we'd brought to be ground the previous week. We drank home-brewed beer from quite a young age and would always go down the cellar for a drink of it…Pub beer would upset us and make our head ache. The farm house had no range but had two brick ovens. In one we kept the powder and cartridges dry – I often wondered what would happen if someone lit a match carelessly. The other oven was used for Monday baking. We'd use surplus faggots bought from Mr. Holden at the Greyhound, burn the wood till the oven was hot enough and then close it down to keep the heat constant. We never used coal.

[There were many people named Holden in the area. Lord Leconfield would quip that there was an 'old 'un' in every wood, meaning a fox, but implying a Holden. Faggots were bundles of twiggy hazel sticks.]

My father made a barometer out of a jam-jar; he filled it two thirds full of water, and then got an old olive oil bottle which he rested upside down on the neck of the jam-jar. The drier it was the more the water would rise into the inverted bottle. We set great store on this and kept it in a little square window of its own, looking at it every morning.

When Lord Leconfield bought the Peachey farms just before the Great War he had the well at Palfrey examined; we found that it was actually inside the house under a great stone slab. He wouldn't buy Ebernoe House itself. It would cost too much to repair to be of any use to him."

Squire William had built a structure on land near the church in 1874 where Mrs. Pullen ran school classes and in 1878 he donated the land to the Diocese for formal recognition as a Church of England 'National School'. The building was 30 ft. by 17 ft. built with local Ebernoe brick, open to the roof with wooden beams and a wood floor.

The 'National School' was funded by the Church of England with pupil payments of a penny or twopence a week, with some grants of government money only if 'results' were deemed satisfactory. Attendance rates and 'standards' were checked by Inspectors. Previously only a Dame School held at Warren House or in the church had offered elementary education. Now in 1877 the school log book began. 68 children were enrolled from 36 families, 29 of whom were labourers by calling. Farming was then labour intensive and the population of villages much denser than today. Contraception was little understood and inevitably unemployment and poverty drove many to emigrate to America and the Colonies. In Victorian times George Wyndham, Earl of Egremont, had actually chartered a ship to take 54 local people, with others, to Canada. The Diocesan Inspector came regularly to check on knowledge of the New Testament and the Catechism taught by Miss Steer and a Monitor. The staple curriculum was the three R's and 'object lessons'. Attendance varied with the weather and occasional demands to pick blackberries, hops, chestnuts, and acorns for the pigs. Pork and pig offal was the staple meat of most households.

Ebernoe School, the earliest known picture

Ebernoe School 2012 as a private house.

In 1870, however, the Education Act at last established free elementary education for all. But facilities were poor. The outdoor privies, two each for boys and girls, were serviced by leaking buckets and any water had to be fetched by wheeled barrow from distant wells. Inadequate heat was dispensed by a smoky 'tortoise' stove. Standards slipped and in 1900 the Inspector described the teaching as 'perfunctory and slovenly'. 'The children need discipline and cannot pick up long division'. Six Wakefords were registered as pupils before the turn of the century. In 1902 an Education Act abolished 'School Boards' and vested education and schools in the Local Education Authority, so that voluntary schools were partly paid for by the rates. The provision of buildings and repairs remained the responsibility of the Diocese. Ebernoe church-goers could barely afford the repairs, but the Vicar, Rev. E. Hamilton Elliott, an occasional teacher, donor of bulls-eyes and a keen tee-totaller steered the school to relative success during George's years as a schoolboy. There was a crisis when £175 was needed to match LEA building standards. Miss Wakeford, George's aunt Gert at Birchwells, wrote to the manager of the Leconfield Estate to support the Vicar's petition objecting to the imminent closure of the school. A reprieve was granted, initially to 1917. The Elims, now of Ebernoe House, raised funds by jumble sales and a concert and the necessary improvements, a boys' urinal, a coalhouse, new stove, lavatory pails and some scripture books were paid for so that the school survived. Classes continued in fact for half a century. The school closed and became a private house in 1951.

Rev. Hamilton Elliott. Vicar 1894-1919

Headteacher, Miss Humphrey, was succeeded by Miss Brows who had 'problems of discipline' and briefly Miss Chislett in early 1907 when George began school, as did Miss Gertrude Bridle. She was there six years providing four years of George's tuition. He left in 1911, one of nine children with the Wakeford surname registered after 1900. She it was who pinned up the bee poster that George ignored. It was recorded in the log book that she had to put up with mice gnawing the school museum and the old registers on top of the cupboard. The Wakefords no doubt subscribed to the electro-plated dinner cruet which was presented to her on her birthday in 1910, a mark of their respect.

Miss Bridle and the class of 1909. George is believed to be 3rd from left, back row.

George undoubtedly benefited from a sound basic elementary education supplemented by his years at Petworth Area School. His subsequent achievements clearly demonstrate his innate ability and intelligence. The West Sussex Gazette reported that the Head of Ebernoe School had sent in a cowslip found by a pupil at Palfrey Farm which had 108 flowers and buds. That pupil might well have been George. The historian is often tempted to ask 'What might have been?' had some former circumstance been different. What if educational opportunities for a bright youngster had been similar when George was born to what was available when he died? He was a schoolboy in an era and social setting which totally proscribed any serious form of secondary or higher education for any but the well-to-do. Had he been born a century later he would surely have climbed the educational ladder to university to become a Professor of Natural History or a reputable environmentalist. As it was, he taught himself in the 'University of Life', by observation and application, to become celebrated in his own life time.

One of my very first memories, I think, was the skep we were given after my Dad had cut the main part of the honey out. There was a little odd piece of honey left in, and we were given the skep. my brother and I, and a spoon each to get busy with what honey we could find in the skep.

I remember when I was about twelve years old my mother came into the bedroom one night and started me on a prayer. She seemed pretty well as nervous as I did, and she got me on to "Gentle Jesus, meek and mild, look upon a little child." I don't remember the rest now.

Another thing I remember was when we would be watching the rain in the afternoon, and sometimes in the evening it would clear up nicely and the sun would shine. So off we used to go, with our heavy boots on, and track the water from one track to another and get it to run off our road which led to the farm. We used to have long sticks and we knew every pipe entrance in the land drains on the farm, and we used to open them well out as far as our sticks would reach.

Another thing we would do was looking round to find our grandfather that was my mother's father, who would perhaps be cutting hedges and cutting out pea and bean sticks for our own use. We would get him to make us a kind of spade out of a nice piece of hazel; we used to cut so many hedgerows every year for our own farm. We made our spars for pegging thatch on. With the spades he would cut out for us we were always willing to dig the docks out of the land. We had very few docks in the end for they were always taken off very carefully to the hedge so that no seed would fall in the field to come up again. We called this thing a "Dig docker". We used to help quite a lot in keeping our land free from docks.

I also remember my Dad's father, Grandfather Wakeford, though I don't remember him very well. We used to go for walks with him. Granny used to live within about half a mile of us on the Common, and when we started school we used to pass quite close to her house in the middle of Ebernoe Common, a place called Birchwells. Also my Auntie Gert lived there; she was my Dad's sister.

I started at Ebernoe School at the age of about seven. I waited until my brother was just on six and then we went together. We had about a mile of mud to go through pretty well when it was winter time if we managed to get down there. The first thing I noticed on going into school was a diagram of a worker bee with all the parts named, but as they were named in Latin they were just no use to me or to any others who went to that school. I don't know where that diagram would be now. It was quite large, about three feet high by two feet wide, I would think. I don't remember a lesson ever being given from it.

I'd like to tell a bit about my Mum and Dad at Palfrey Farm. They took the farm in about 1900. They had no means of cutting corn only with a scythe or a faghook for about four years. I can remember the first selfbinder that was ever used on Palfrey Farm. I think it was a Hornsby; they were among the first to be had, but until then my Dad had to cut all his corn with a scythe or a faghook, and tie it up with the straw of the corn. I don't know how they would get on these days with all that lot. They did manage. After that with the reapers and binders, my Dad generally used to get his corn cut by someone, and in the end we had a brand new Binder, a Deering Binder, which did some jolly good work for us.

The American Company, McCormick introduced the horse-drawn Deering selfbinder with a knotting device to bind the sheaves in the 1890s. The Hornsby was a British development. Earlier 'sailor' reapers had cut the corn and swept it into bundles which had to be tied by hand. A man with a scythe had been expected to cut only half an acre in a long day. Each new reaper saved the labour of six men and a tier. Despite this economy of labour, farmers struggled to compete with the price of imported grain also coming from the New World. Unemployment which resulted, contributed to depopulation from emigration to the towns and overseas.

My Mum and Dad were very close and worked well together in all farm work. My mother had a job mostly for every day of the week. One day was for baking when a large oven was heated and filled with three or four large loaves, three or four large cakes, rice and other milk puddings, rabbit pies, buns and rolls. She did this once a week. There was a regular day for butter-making. She'd make up to thirty pounds of butter all in half pounds. In the first place we had little old box churns that sometimes took the whole day before we could get the butter. Later on we had a Bowby two-minute churn made by Robert Bowby of Bury St. Edmunds.

[Robert Boby founded his business in Bury in 1843, manufacturing milling, malting and other food-processing equipment. The company was taken over by Vickers in 1927 and the works closed in 1970.]

With this when you turned the outside handle once the inside went around three or even nine times, and many the times we got the butter in about two minutes which was a great help, but my mother couldn't do that on her own. We boys used to do that. When the butter first comes there's still an amount of milk left in the butter. This had to be worked out by continually mucking the butter about. Finally it's put out into a big wooden bowl ready scalded so the butter couldn't stick to it and tapped with a cloth or a cheese cloth to collect any further moisture.

When this is done thoroughly it's slightly salted and that's all worked in. It's made into half-pound pats, similar to what you buy now only ours was nicely printed on top. It was my mother's glory to cover it over after she'd finished and take it away to the dairy. This was an underground place which let a little light in from above. It was mainly three to four foot underground, the same as the double cellar adjoining.

Chapter II

LIFE AT EBERNOE

A well-preserved coat and hat with a small skin in the wearer's hand.
200 or more skins would be used.

We left Ebernoe School when I was about eleven, but not before we'd caught quite a lot of moles to make our head governess [the local term for the Headmistress, Miss Bridle] a coat. This all happened suddenly this governess decided she would like a coat out of it. Other boys beside me started catching moles, but I took on to dress all the skins, and I used to make a solution of alum and saltpetre. I tacked the skins out on the board and dried them first, then I used to brush them over with the alum and saltpetre each night. She was quite a big person and I can well remember her in the Ebernoe church with this coat on and it almost reached the ground. I can't say how many skins it took to make that coat, but quite a lot.

Whilst I was there on one occasion we had a bit of snow. When we were out for play, you could get quite a long way from the school on the Common, but you didn't really want to go too far from the school for when the bell went. One day I was quite a distance away when we had. this snowstorm and I was dashing back when one of my school mates jumped out from behind a bush and aimed a snowball at me and caught me smack on the nose. My nose bled very freely and they got me in the school and had me flat on my back on one of the desks. I think some of them were frightened rather more than I was.

We used to take our dinner to the keeper's house fairly close to the school, leave the bags there and go on to school and then back with the two sons of that house, name of Turner. Mrs. Turner had our food on two plates and we used to sit in the chimney corner to eat it. They were a good family. Mr. Turner was keeper for Squire Peachey, the owner of a small estate and he also owned the farm. They used to come round pheasant shooting at times and I happened to get in with the beaters. At Palfrey we had rather a pretty pheasant. He was called a pranked pheasant. He had a good many white feathers in him. As we were beating in Palfrey Copse, that's the copse right close to Palfrey Farm, in the first drive a hen pheasant got in right against Mr. Turner who at that time was carrying a gun and one of the local beaters said, "Now George, now George," and Mr. Turner shot at it. He knocked some feathers out, but the bird kept going. On the next drive I was beating through and in line with the rest and I looked in the bush and there was this pranked cock pheasant. I said to the others that he was in there. "Muck him out, muck him out," they said. I thrashed the bushes and out he went and we hollered "pheasant coming over". They had several shots at it but they didn't get him.

The old shoe-mender, a man by the name of Jimmy Spooner lived at Balls Cross. He then shifted to his friend's, right on the cricket ground at Ebernoe, and he did his work in a caravan. One day I went there for some mending, I picked it up and there was a set of geese there. The old gander followed me all the way across the cricket ground but he didn't attack me or anything. I just went backwards all the time. I think the tale got about that he got on my back, but he didn't. He just kept hissing and following me all the way.

I remember Mr. Spooner at the first cricket match I played in there. It was a schoolboys' cricket match against Kirdford. We lost pretty heavily. The old man, Mr. Spooner, who was quite a character, umpired for us. On that ground the Horn's Fair is held every year when they roast a sheep on the 25th July, unless it falls on a Sunday. If it does the fair is held on the Saturday. It is named Horn Fair because there is a horned sheep roasted for lunch there every year. Throughout the war years we couldn't have a sheep roasted for lunch so the Leconfield Estate supplied us with a deer.

The highest scorer on the winning side wins the horns of the sheep. I never won it myself, but I was runner-up on one occasion. We used to have a way with Dad. He would go to the Fair until sometime in the afternoon and then he would come home and tend to the cattle and that, whilst we boys went from then on in the evening. We always wanted to get down there to see Les Blakeley batting and also we always wanted to know if Dan Pennicott had been in. He was a demon bowler. He could spin the ball in from each way and throw it high in the air straight up over his head and I never saw another one do that at any time.

George was a keen cricketer, nick-named bucket hands for his catching skill. Derek Adams who played with him for Wisborough Green C.C. says he would rattle a cricket ball round in his great cupped hands. Once when they were fielding at Kirdford he saw George run off the pitch to gather a swarm while the rest of the players threw themselves on the turf. A friend once asked him to expel a mouse from a hive on the Maxse fruit farm in Fittleworth. At the first attempt the mouse streaked out and escaped into the bushes. At the second George took up a cricket fielder's stance and shouted to his friend to strike the hive with a stick. Out jumped the mouse from the front entrance and was firmly held in the cricketer's grasp.

Cricket team (1926) showing Edgar Fiest (back 4th left), George (5th left), Rev. Standish (8th) and David Baker (Front 2nd from right).

The term Horn Fair on the face of it is named because of the prize, but country folk knew that a racket from cows' horns, the clatter of metal tools and other 'rough musical' uproar, was the traditional way scandalized villagers took action to remonstrate with a wife or child beater or anyone who indulged in other indecent or anti-social behaviour. Horns were also worn, time out of mind, to mock cuckolds and adulterers and to accompany all kinds of lewd ribaldry- what Chaucer described as 'syne and harlotrie'. The substitution of a cricket prize, a good meal and dancing would have been an innocuous Victorian substitute for lewd behaviour and 'robust humour'. Ralph Vaughan Williams collected the Horn Fair folk song in 1904, when George was four. Here are extracts from the lyrics suitably suggestive of earthy rustic romance:-

A I was a-walking one mid-summers morn
So soft was the wind and the waves on the corn.
I met a fair maid upon a grey mare
And she was a-riding along to Horn Fair.

"Now take me up behind you, fair maid, for to ride."
"Oh no and then no, for my mummy would chide,
And then my dear daddy would beat me full sore,
And never let me ride on my grey mare no more."

"I can see by your looks, you've one game of play,
And you will not ride me nor my grey mare today."

The singer's game certainly wasn't cricket!

It is possible that the song was sung to the same tune as the bawdy ballad '*The threshing machine*'

I once knew a farmer, I knew him quite well'
And he had a daughter, her name it was Nell.
Although she were only the age of sixteen,
She wanted to see my old threshing machine! Etc.

Horned Sheep, roasted whole on a spit on gorse bedecked common

Horns in the pavilion at Ebernoe

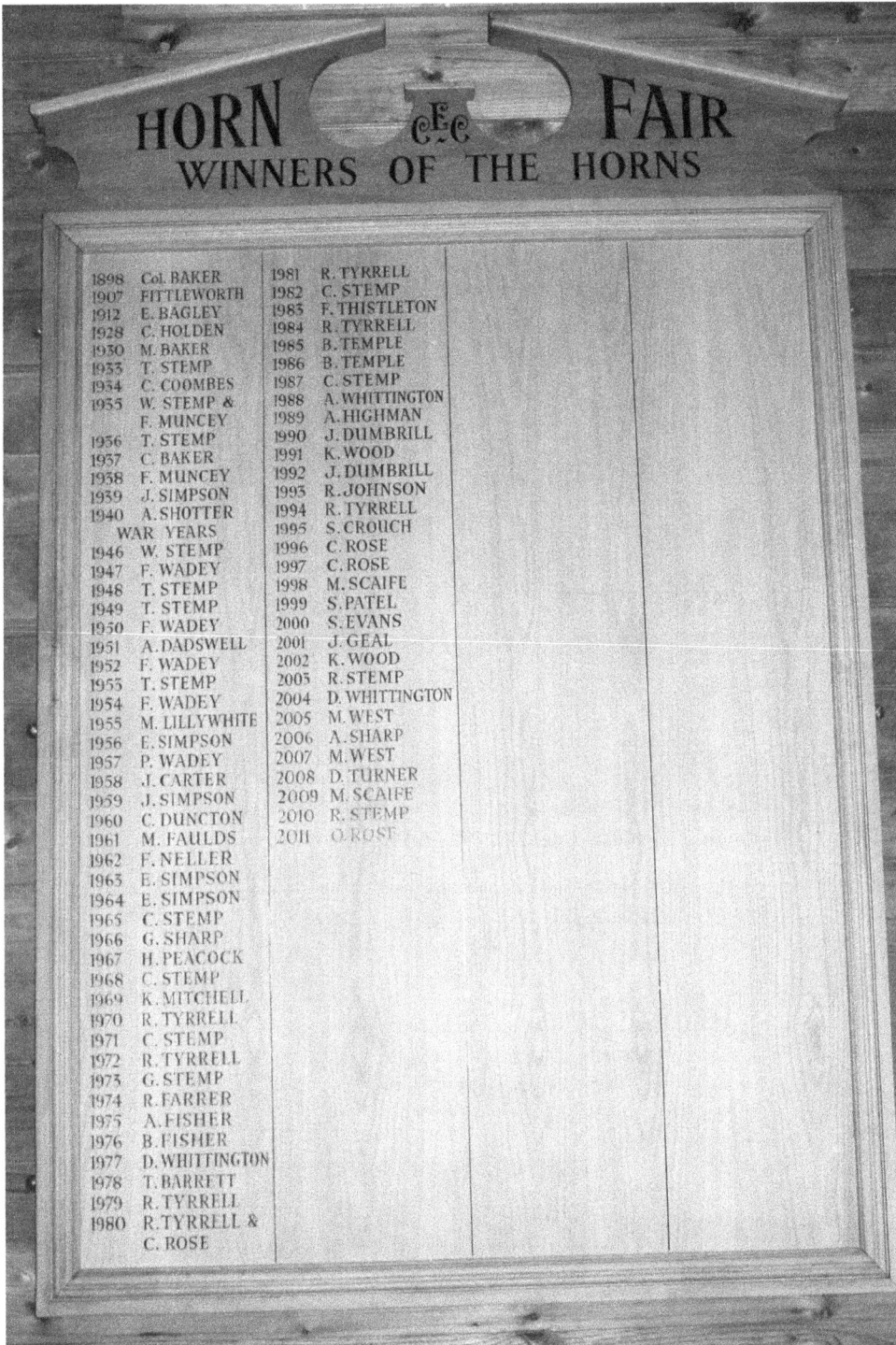

HORN c&c FAIR

WINNERS OF THE HORNS

Year	Winner	Year	Winner
1898	Col. BAKER	1981	R. TYRRELL
1907	FITTLEWORTH	1982	C. STEMP
1912	E. BAGLEY	1983	F. THISTLETON
1928	C. HOLDEN	1984	R. TYRRELL
1930	M. BAKER	1985	B. TEMPLE
1933	T. STEMP	1986	B. TEMPLE
1934	C. COOMBES	1987	C. STEMP
1935	W. STEMP & F. MUNCEY	1988	A. WHITTINGTON
1936	T. STEMP	1989	A. HIGHMAN
1937	C. BAKER	1990	J. DUMBRILL
1938	F. MUNCEY	1991	K. WOOD
1939	J. SIMPSON	1992	J. DUMBRILL
1940	A. SHOTTER	1993	R. JOHNSON
	WAR YEARS	1994	R. TYRRELL
1946	W. STEMP	1995	S. CROUCH
1947	F. WADEY	1996	C. ROSE
1948	T. STEMP	1997	C. ROSE
1949	T. STEMP	1998	M. SCAIFE
1950	F. WADEY	1999	S. PATEL
1951	A. DADSWELL	2000	S. EVANS
1952	F. WADEY	2001	J. GEAL
1953	T. STEMP	2002	K. WOOD
1954	F. WADEY	2003	R. STEMP
1955	M. LILLYWHITE	2004	D. WHITTINGTON
1956	E. SIMPSON	2005	M. WEST
1957	P. WADEY	2006	A. SHARP
1958	J. CARTER	2007	M. WEST
1959	J. SIMPSON	2008	D. TURNER
1960	C. DUNCTON	2009	M. SCAIFE
1961	M. FAULDS	2010	R. STEMP
1962	F. NELLER	2011	C. ROSE
1963	E. SIMPSON		
1964	E. SIMPSON		
1965	C. STEMP		
1966	G. SHARP		
1967	H. PEACOCK		
1968	C. STEMP		
1969	K. MITCHELL		
1970	R. TYRRELL		
1971	C. STEMP		
1972	R. TYRRELL		
1973	G. STEMP		
1974	R. FARRER		
1975	A. FISHER		
1976	B. FISHER		
1977	D. WHITTINGTON		
1978	T. BARRETT		
1979	R. TYRRELL		
1980	R. TYRRELL & C. ROSE		

Honours Board in Ebernoe Pavilion

(4 miles north of Petworth, A283)
A Grand Old Sussex Custom

EBERNOE HORN FAIR

Monday 25th July 2011

-8.30am-	-11.00am-
Sheep Roast Commences	Ploughman's Lunches
	Afternoon Teas available
-11.00am-	-6.05pm-
Cricket Match Commences	Presentation of Horns
Ebernoe v Lurgashall	Children's Sports
	Sheep Roast and Barbecue

**GRAND TOMBOLA – SPIRIT BONANZA
DIP THE LADY – LICENSED BAR 11am to 11pm
TRADITIONAL SKITTLES & STOOLBALL DRAW
PRIZE DRAWS - EXCITING FUN FAIR
PETWORTH TOWN BAND
& MANY OTHER ATTRACTIONS**
All communications to: 01798 344590 / 07799 246806

Notice of Horn Fair 2011

The re-sited pavilion today. David Wakeford paid £90 towards it and £100 was raised by subscription

When we left Ebernoe school we went on to Petworth School, a bigger school altogether. We had about three miles to go. We walked it for about two or three years and then my Uncle Steve managed to buy an old bike for each of us. They were old ones too, but we managed quite nicely to get to school on them in the morning. That was all pretty straightforward, but when night came some of the lads coming our way liked to hang on the back and then we didn't get home quite so soon. I got on well with all the teachers there except one who was rather unpleasant.

In September 1942, a Heinkel 111 bomber dropped three bombs, one of which hit Petworth Boys School, resulting in the death of 28 of the 80 pupils present, the headmaster, an assistant teacher and another adult. Many more were injured. Canadian soldiers, encamped locally, rushed to the rescue and assisted in the subsequent burial in a mass grave. One of the children was Alan Simmonds, like George a former scholar at Ebernoe School.

I left the Petworth School when I was about fourteen. It is worth recording that even in those early days I had found one way of earning an honest penny. Most time travelling to and from school we used to catch and then skin moles. We were able to get a regular 3s. postal order for every twelve moleskins sent to London. With this we could buy a hundred three-bore cartridges for two-and-six and a sixpennyworth of sweets.

The Vicar at this time was the Reverend Hamilton Elliott, a very respected man in the village. His wife played the organ. My brother and an Ebernoe schoolgirl and myself went to the Vicarage for Confirmation classes. One day when we were sitting around the table he tossed his pencil down and asked me if I believed that it was on the table. "Yes I do." "You don't believe it's on the table. You know it's on the table" - pointing out the difference between knowing and believing. He would have to be wide awake now or I might catch him out with something similar.

On leaving school I wanted to go gardening, and to ask for a job at Ebernoe House. Squire Peachey was living there at that time. He had two gardeners, Mr. Feist and Mr. Baker. Mr. Feist was the head gardener and he was a very spruce clean-living chap who went to church every Sunday morning. He was a left-handed cricketer and a good cricketer at that. He didn't bowl or anything much, but he usually kept wicket.

David Baker was another one of the cricket team. He was a good little bowler, and that was mainly his job. I really wanted to start gardening under them; but my Dad said he would give me a job right away, and he started me on the farm. I think he would have been very hurt if I had not joined him. I suppose I didn't make a very early start the first morning, because he said he wanted me a little earlier the next morning. I was not going to have this said, and when I heard him getting up in the morning, I was always ready to follow him down the stairs.

Before I left school I wanted a small-bore gun and asked my Dad for one. He said, "You can have a gun when you start doing some work." This proper got us going and we soon had the little gun we wanted. My Dad was an excellent shot and finally all three of us were about as good.

> [The small-bore gun was probably a .410 calibre shotgun which used a small cheaper cartridge and had lower recoil than a 20 or 12 bore. It could have been a .22 rifle. The .410 was suitable for boys and mainly used for shooting rats, rabbits and other vermin. George wrote of a 'three bore', an unfamiliar description.]

He was pleased with the farm he had worked so hard for. He understood the land and was a good farmer. I was mainly with the cows and when my brother left school, he took over the horses and ploughing. We had a two-wheel plough and my brother was doing well with all but the ridge of the ploughing. This is the centre where you start the ploughing. You go down one way and turn some over and then another piece over and then the next piece back. My Dad went with him and showed him just where to let one of the wheels go. When he went again a few days later and my brother was carrying on, my Dad said, "You can now plough as well as I can." My brother soon made an excellent ploughman and won many prizes at local ploughing matches.

George would have been an early riser as he worked with the cows and would have to help with milking by hand, squatting on a three-legged stool, in the early morning. Butter was their main product, there being no local arrangement for the collection of raw milk at the time. One hand-milker was needed for every ten cows in a dairy herd.

We worked well and liked all farming jobs. It was six days a week work. On Sunday we only did the necessary work that led to the animals' comfort, and we were fairly good attendants at Ebernoe

Church. When I was about eighteen or nineteen my Dad was ailing and my brother and I had all to do. We would plough and sow a first plot of land and then ask him if we could start on another one. Then we cleared and sowed, mainly wheat it was then, the next plot, and so we would go on. He died that year and then we had it all to ourselves, my mother, brother and me, but we managed to get through the work.

Ebernoe Church today

Interior View of Ebernoe Church

John Peachey died in March 1911. His considerable estates in Gloucester had already been sold a decade earlier, realizing £16,000, but after clearing William's legacies little remained and the West Country rental income vanished, leaving only the meagre Ebernoe rents to live on. The estate had to be sold. At auction it failed to reach a £17,000 reserve. Lord Leconfield offered £14,000 plus £10,000 for the timber and eventually settled for £29,000 without the house and 15 acres, so safeguarding his borders against outsiders and any threat to his hunting rights. His agent, J.B. Watson's report of June 1911 reads: 'Looked over Palfrey Farm. The arable land is for the most part clean and well-farmed by Mr. Geo. Wakeford. There are 3 very rough fields of pasture – probably 'fallen' to grass some 14 or 15 years or so and worth little at present. 3 or 4 fields require draining. The buildings and house are in very poor condition. Considerable outlay is required. The rent of £40 is very moderate for 107 acres and would be a fair basis for purchase. J.B.W'. Repairs there cost him £164!

The Elins, London business people, bought Ebernoe House. Mrs. Elin had been on the stage and astonished local opinion by wearing trousers to tend the garden. The contents, furniture and implements, including a well-varnished dogcart, extensive beer-making equipment and seven hound beds, were sold off at auction in a 3 day sale in 1912. Many believed that much beer had once been brewed and sold on the premises as though it were an inn.

My Dad also taught me to make rabbit nets. He'd never made one in his life, but he used to see his father doing it and he had a good idea from that.

Boys ferreting at the turn of the century

Purse nets would be placed over all the holes exiting a 'bury' or set of underground burrows. The rabbits would then be driven out by a ferret and trapped in the nets, each of which nets had a drawcord anchored with a peg.

His father used to be keeper for Lord Leconfield and at that time he used to fish the ponds about every four years and catch and take away the biggest of the fish and put the smallest ones back again. They used to let all the water down a ditch. They had a proper pen stock to pull out to let the water out of the ponds and then they used to put it back and let the water in and stock up with another lot of fish. My grandfather used to make these nets to let the water go through but so that all the fish didn't get away.

Whilst talking to the Vicar of Kirdford, Mr. Newman, recently about all the changes that have taken place in the last seventy years we marvelled at the progress during this period. I was taking honey from the bees in a hive whilst the bees were still in it while my Dad was unable to do this with his skeps unless he first killed the bees.

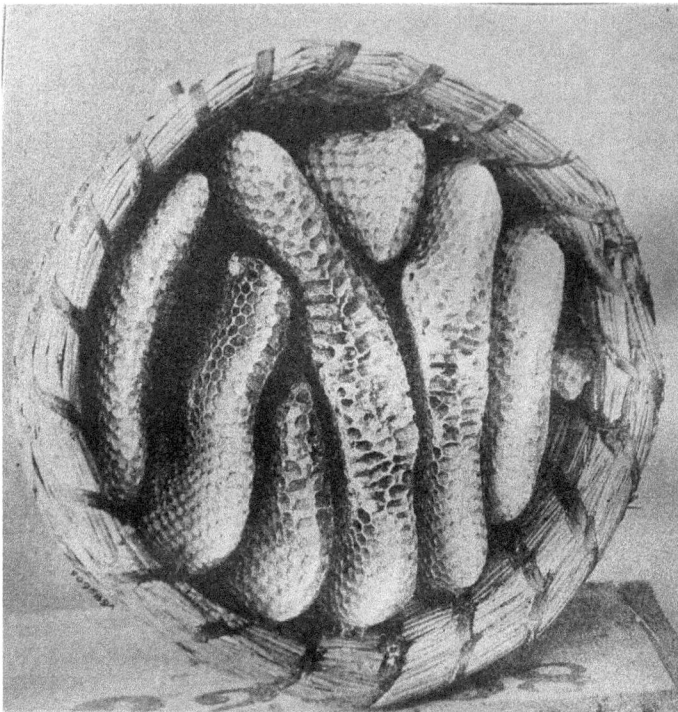

Combs built inside a skep

ARISTAEUS, THE FIRST BEE-MASTER
(From Floris 1435-1570)

Aristeus, First Bee-master, mythological Greek god, the 'Inventor of Beekeeping' with skeps

Skep hive in a bee-bole

Skep hives under milk pans

George was quoted elsewhere as follows:

'Probably my grandfather kept bees when he was at Frith Hill near Northchapel but when I asked my Aunt Gert about it she just laughed and said, 'I think he was afraid of them'.

'Father kept bees at Palfrey and always had skeps covered in sheeting by the east wall. He didn't sell the honey. A large number of the older people kept bees. They treated them as a natural crop like field mushrooms; often they simply used old barrels instead of skeps. There were blackberries everywhere then, more wild flowers and of course no spraying. The skeps were made from straw by a family at Fittleworth who would walk round the outlying farms to sell them. The skeps were quite light and they would be carried, tied together, and slung over the shoulder. My father never made skeps himself. I never saw a skep made of anything but straw. This hive had a wooden band two or three inches in width at the bottom.

My father was in no way a scientific beekeeper: he couldn't tell a worker from a drone nor pick the queen. He knew simply that bees needed a good hive that was sound, clean and dry. Beyond that he was prepared to leave the bees to their own devices.

The three genders of bees that George knew and his father did not:-

Queen and Workers

Drone

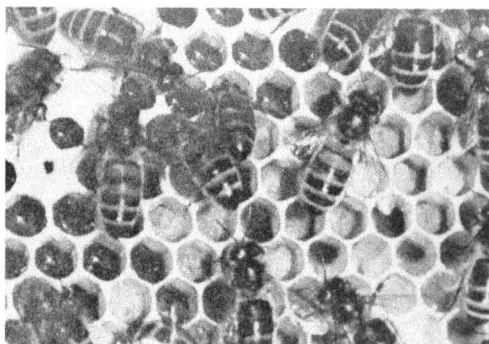

Worker bees

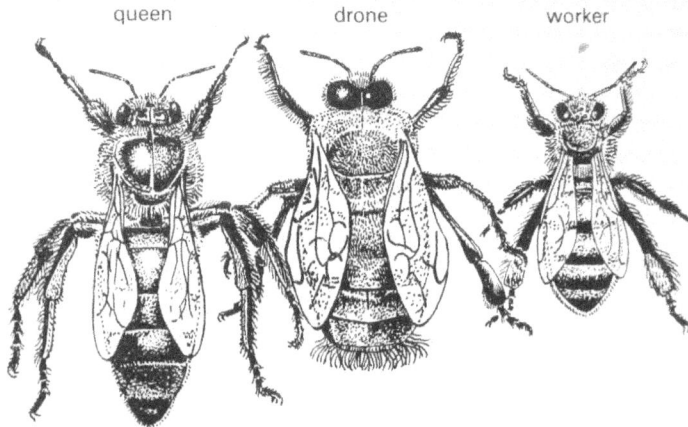

| queen | drone | worker |

the shapes to look for

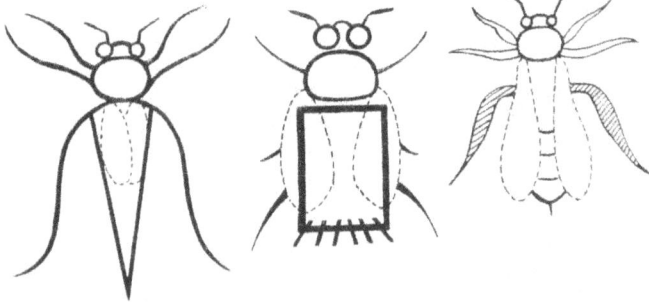

The three castes of bee

| The 'widowed' egg-laying mother of a single-parent family of maybe 15,000 workers and 200 drones. | A male bee, one of several 'sperm donors' with just one mission, to fly, accost a young flying queen, mate and die. | A sterile female bee, nurse, forager, wax comb builder and hive guardian. |

Every beekeeper would leave a couple of hives to overwinter. My father was no exception. If he had over-wintered his colony successfully his year would effectively begin... when spring turned to early summer and the bees were building up to swarm. When they did swarm he'd dress up in protective clothing ...and go to collect them. You couldn't buy protective clothing in those days; you'd simply made it yourself. He'd dress up in an old curtain and cover his hands with old socks. When he was fully equipped I always thought it was a miracle he could see the bees at all, let alone take them...He would put a sack on the ground, place the skep on the top of it and carefully shake the bees into the skep. The skep was solid at the top and needed to be set back at an angle with a brick or something similar; this angled opening gave the only ventilation...It was all too easy to suffocate even the strongest swarm... I have done it myself. Some older farms still had bee-boles in the walls to hold the skeps but we simply placed them on a low stand.

[Two houses in Petworth, Denne Court and Boles House, both in East St. have bee-boles according to the International Bee Research Association.]

... He would hive them on the spot and in the evening at dusk when every bee had returned he would carry the skep to its permanent position. The skep or wooden hive would be covered in sacks for warmth. A broken bread pan was ideal to keep off the rain.

There were a considerable number of wild bees about then and if a swarm came across the farm my father would chase it. Yes, they did use the old method of 'tanging', banging on a shovel with a key to arrest the swarm, although I never had much confidence in it.....

Old print depicting the tanging of a swarm

If he did not take the swarm properly he would bruise nettles and lay them where the swarm was first taken because they would tend to return to that place. I never found it very successful...You rarely see wild swarms now....

In those days you could only take off one crop of honey – in August/September.' The night before taking the crop he would separate the hives from their stands.... they would be stuck fast... He would then feel the weight of the skeps and chose the heaviest two to overwinter – he knew nothing about feeding and they would simply live off the honey they had made. He had no smoker to subdue the bees so that they had to be killed to get at the honey. The following day he would dig a single pit in front of the hives, big enough to take a skep. Into this he'd put a brimstone stick-we used to buy them from Mrs Knight's shop at Petworth and knew them as 'matches'. A match looked like a piece of corrugated brown paper. When it was lit and smoking he'd take a hive with bees in it and place it over the burning brimstone. Within two minutes the bees would be suffocated. If the hive was very strong there would be three or four hundred left running on the stand. These he would cover with a sack and crush by pushing down with his hands. He would do all the hives in a single evening.'

The skeps would be taken to the kitchen and the combs cut out or simply broken off using the weight of the honey as a lever. The brood and developing bees and pollen would be removed. Then the combs would be crushed. Usually it was pressed with a knife until the honey ran out into one of the old three-pound jam jars. Over the winter it would gradually crystallise – a sign of good honey. My mother let the honey seep through muslin bags and used the combs first to make mead and then to make home-brewed beer. The combs would then be melted into a Keiller's stone jar; a little honey and vinegar were added and constantly stirred until it formed a paste. This 'beeswax' was used as a polish and would keep for years'. My mother would never cook with honey; sugar does the job of sweetening rather better. All our honey we ate on bread. Mead, if made properly was very strong and considered something rather special compared with the other home-made beers and wines.

When I went to school the first thing I noticed was the big printed drawing showing the parts of a bee and giving them their technical names...How often my attention wandered during lessons....The teachers never mentioned it – perhaps they were as perplexed by the Latin names as I was.

I first saw modern-type hives when I was about 20. Mr. Purser at Hampers Green had them and he induced me to join the Sussex Beekeepers' Association. Jimmy Baker of Blackwool Farm lent me a book on modern beekeeping and Mr. Kenward was employed by the SBA to go round advising people on the new methods. Every member could receive a visit if they so wished....My father thought the price of the new hives was exorbitant. They were over £2 each. I did, however, invest in the new hives and began to keep bees on a scientific basis.'

Kind-hearted beekeepers had long regretted the necessity of killing the bees in order to take their honey. During the previous two centuries efforts were made to save the problem of 'humanity to honeybees' as it was called. 'The wilful destruction of bees is so wantonly inhuman and cruel that we can scarcely find words to express our abhorrence of the practice,' fulminated the Editor of the British Bee Journal in 1873. 'The bee-cide felon, called man, digs a pit, lights four ounces of brimstone inside of it, and deliberately sits 15,000 bees, queen and all, above its really and truly infernal fumes and burns the unhappy martyrs and then subscribes to various charities in winter and calls himself a philanthropist! He ought to be sent to the treadmill!'

Efforts had been made, successfully but expensively, to overcome the problem by devising wooden hive boxes which offered space for bees to store honey in removable containers above or at the side of the main brood box where the young bees were reared. Such equipment was beyond the means of most village cottagers, who, like George's father, stuck to the old ways. Some even argued that the newer box hives, which allowed all the bees to overwinter, would encourage disease. Another way of saving the bees was to render them senseless by smoking them with dried puffball as fuel but was little used by country folk.

Thomas Hardy's novel *Under the Greenwood Tree* was first published in 1872, just one year before Charles Abbott, a London Beemaster, launched the *British Bee Journal*, which publication led to a revolution in late Victorian beekeeping. During the following thirty years, apiarists learned how to use removable frame hives instead of simple boxes or skeps in which to keep their colonies of bees. This change not only increased yields of honey but also meant that the bees did not have to be suffocated or anaesthetised in the autumn to allow the beekeeper access to his crop. Thus he had many more healthy stocks to gather nectar the following year.

Hardy was but one of many intellectuals who deplored the steady deterioration of traditional village culture which was precipitated by depopulation through migration to towns and the colonies, by agricultural depression and by the introduction of steam-powered implements. Rural dereliction was similarly evident at Ebernoe. In this novel he symbolises such cultural loss through the supersedure of the loyal band of friends in the Mellstock Church choir by the efficient new church organ, to be played by a modern, clever, pretty and somewhat worldly young schoolmistress, Fancy Day.

 Hardy's novel is sub-divided into the seasons, tracing the courtship of Fancy by Dick Dewey. In spring ,summer and autumn scenes, there are brief passages depicting the old-fashioned style of beecraft which would have been standard practice in the earlier years when the novel was set. These ways were those familiar to George Wakeford when he was a young boy. George was born at the turn of the century, his boyhood days passing neatly within the first decade.. Hardy was 60 when George was born and no doubt would have been well aware of the new beekeeping methods but his story accurately echoes and reinforces George's record of his earliest experiences with bees.

In summer Dick is told by his father to take two hives of bees in the best spring-cart some ten miles off to the Vicar's mother, 'who had just taken it into her head a fancy for keeping bees (pleasantly disguised under the pretence of its being an economical wish to produce her own honey)'. In those days the clergy were great advocates of beekeeping.

In autumn there is a chapter headed 'Honey Taking', with a full account of the 'taking up' of hives.

By the light of the moon and a lantern a procession approached a row of beehives. Enoch the gamekeeper's trapper had brought a spade and Mrs. Day was carrying 'curious objects about a foot long in the form of Latin crosses (made of lath and brown paper dipped in brimstone - called matches by bee-masters).' Mr. Day took the spade and dug two holes in the earth beside the hives.

'The preliminaries of execution were arranged, the matches fixed, the stake kindled, the two hives placed over the two holes and the earth stopped round the edges.

"They were a peculiar family," said Mr. Shiner, regarding the hives reflectively. Geoffrey Day nodded.

"These holes will be the grave of thousands!" said Fancy. "I think it is rather a cruel thing to do."

Her father shook his head. 'No,' he said, tapping the hives to take the dead bees from their cells, 'if you suffocate 'em this way, they only die once: if you fumigate 'em [by burning puff-ball] in the new way, they come to life again and die o' starvation; so the pangs o' death be twice upon 'em."

"I incline to Fancy's notion," said Mr Shiner, laughing lightly. [He too was in love with her].

"The proper way to take honey, so that the bees be neither starved nor murdered, is a puzzling matter," said the keeper steadily.

"I should like never to take it from them," said Fancy.

"But 'tis the money," said Enoch musingly. "For without money man is a shadder!"

The story continues with the takers-up attacked by bees whose hives had been destroyed some days earlier and were now 'getting a living as marauders about the doors of other hives.' [George's father avoided this problem by killing the 'stragglers' that had escaped the fumes by crushing them under a sack.]

Enoch threw down the lantern and ran off and pushed his head into a currant bush. Mr. Shiner floundered away helter-skelter among the cabbages.

'Are those all of them, father? said Fancy, when Geoffrey had pulled away five.

'They can't sting me many more times more, poor things, for they must be getting

weak'.
'Is it all quite safe again? said Mr. Shiner from the bushes.

In Geoffrey Day's storehouse were several empty skeps above barrels of new cider of the first crop, each bubbling and squirting forth from the yet open bung-hole. Fancy is pictured with a hive in her lap 'for convenience in operating on the contents' so beginning the traditional extracting procedure.

'She thrust her sleeves above her elbows and inserted her small pink hand edgewise between each white lobe of honeycomb, performing the act so adroitly and gently as not to unseal a single cell. Then cracking the piece off at the crown of the hive by a slight backward and forward movement, she lifted each portion as it was loosened into a large blue platter.'

Such is the precision of this account that it is clearly a familiar experience of Thomas Hardy in his Wessex youth and touches on the dilemma of what indeed was true 'humanity to honeybees'.
At the end of the novel Dick was almost late for his wedding to Fancy. Time and tide and a swarm of bees wait for no man.

'The hive 'o bees his mother gie'd en for his new garden swarmed jist as he was starting, and he said, "I can't afford to lose a stock o' bees, though I fain would, and Fancy wouldn'd wish it on any account." 'So he just stopped to ting to 'em and shake 'em.' [He still believed a swarm could be settled by 'tanging' or beating a metal object.]
'A genuine wise man,' said Geoffrey.
'That my bees should ha' swarmed just then, of all times and seasons!' continued Dick 'and 'tis a fine swarm, too: I haven't seen such a fine swarm for these ten years.'
'A excellent sign, said Mrs. Penny.

Such was the fashion of beekeeping, as was understood by Mr. George Wakeford when a boy when he might well have read Hardy's novel. All changed for George when he met Mr. Kenward. He was the Sussex Expert, pictured in William Herrod-Hempsall's book '*Beekeeping New and Old*'. His mission was to enlighten those benighted beekeepers who had been described as 'old fogey's' by the Father of Modern Beekeeping, the Rev. Lorenzo Langstroth, a good half century earlier.

Mr. Kenward, County Expert Beekeeper.

Mr. WBC [centre] and Mr. Herrod-Hempsall [right]

In the United States general adoption of box hives, originally designed by Rev. Langstroth in 1851, had caused developments in honey production on an industrial scale. When George was born leading lights in beekeeping such as Thomas Cowan, who lived twenty years at Horsham, William Broughton Carr and the Herrod-Hempsall brothers had so influenced opinion through books, journals, shows and demonstrations by County appointed experts and the adoption by the British

Beekeeping Association of regional clubs, that growing numbers were adopting the new methods. George, by the time he was 21, was cast as a local pioneer of the latest progressive methods. Carr's hive design, the handsome double-walled structure normally recognized by the layman as a 'proper beehive', known as a WBC and still in fairly common use among hobbyists, was his original choice. In dealing with his many clients he dealt with other hive types, in particular the British National, single-walled design and the inexpensive Cottage Hive. He continued using the WBC throughout his life. This design was more trouble to take apart and move from place to place but had the great advantage that the frames in its boxes were British Standard size as established since 1882. These were interchangeable with those used in the National hive, a model increasingly adopted among the vast majority of the amateur beekeepers who were George's clients. Naturally he also utilised all the accessories that went with the modern methods such as factory-made comb foundation and precision-cut frames and honey extractors which whirled out the honey from the combs by centrifugal force. As a beekeeping instructor he would have been aware of the then up-to-date current best practice as popularised by such popular manuals as *'Modern Beekeeping'*, published by the BBKA, Cowan's *'British Bee-keeper's Guide Book'* and the work of the Irish expert, Rev Joseph Digges. However his knowledge came by word of mouth from beekeeping friends and colleagues he met in his work or at shows and at meetings of the local Association. He always claimed not to have read any beekeeping instruction manuals though he harboured a fancy to write one and kept himself informed through *Bee Craft* and other journals. .

Exploded view of a WBC hive

Vintage WBC hives built by C.T.Overton & Sons, of Crawley, Sussex.
The one on the right sold at auction for £300 in 2012

George behind a WBC

The British National single-walled hive

Smoking the 'supers' in a WBC with 'lifts' removed

We also talked about farming progress. For hundreds of years there was little change and in the last hundred years tractors have taken the place of horses. Eighty years ago corn was cut by hand, shocked to dry, carted to the stack, thatched and, as needed, the grain was carried to the barn. Then it was thrashed with a flail, which was made of two sticks hitched together, and then winnowed by a winnowing machine to separate the corn from the chaff. Nowadays there are many varieties of corn available and many roots and grasses. There are new and better manures and better knowledge of how to use them. In the last few years various forms of sprays have helped a great deal in controlling weeds.

In the early days I used to hear a lot about the mighty atom, and I wondered if I would live to see it harnessed. I also thought that the growing of tobacco was a waste of good land, and I hoped, and still do, that corn would take the place of tobacco. The atom has now been harnessed so progress is being made.

Thrashing

The thrashing drum was a very heavy affair and had to be hauled by horses into position. The steam engine that drove the drum was a portable one and quite heavy, and this also had to be hauled by horses. There was a straw elevator which collected the thrashed straw and elevated to the straw stack. In later years another machine was attached to the threshing drum and this packed the straw and tied it into trusses. Later the engine was a traction engine and used to haul the threshing drum, the elevator and the straw trusses on its own.

Now one hardly ever hears the hum of the threshing machine. Looking back to the days of the self-binder I can remember the perfect sheaves of corn that the machines made, far more perfect than any that were otherwise made. When cutting corn with the horses and binder we always liked to finish one field or piece of corn each day because there was almost sure to be some rabbits in the corn. We then had a shoot, getting most of them, which kept their numbers down and they were good in the pot. Occasionally there might be a fox or a hare in the corn, but they would run out of the corn quite early. We did not kill hares as our landlord, Squire Peachey of Ebernoe House, had a pack of hare hounds and we were quite pleased to save any hares that were on the farm.

My Dad died when I was eighteen, that is fifty-eight years ago, and I cannot remember all about him beyond the fact that he was a good farmer, a good rabbit shot and good in every job on the farm. He died at night and although Ebernoe church was a mile away, I heard the knell being rung for him. We seldom heard this bell at that distance. Why should I hear it that morning? It woke me up and I still remember it.

Tree felling

At twenty-five years of age I left home and started a job of tree-felling. This was hard manual work, a little different from what I was used to. My mother and brother did not think I would stick to it, but I got on well with it and was paid £2 10s. the first week. This seemed a handful of money to me, and I put it on the table when I got home and told my mother to take what she liked of it for my keep, but she would not take any of it.

[Mrs. Jane Wakeford, George's mother, lived with Wynne, his younger brother at Hazel House, Elsted after they left Palfrey Farm and George lived with them till he married.]

Team work with a cross-cut saw

I was with men who liked their pints of beer at weekends, but were always willing to work hard for extra money. I was not long at £2 10 shillings [50p], before it was £3 and finally £3 10s., which was the top price paid to any of us. Before leaving the farm I had tried myself out with an axe and cross-cut saw; I got on well with the axe, but not so well with the saw, so that when it came to me to take my place to saw a tree down I wondered how I would get on. I found it quite easy as their saws each were well sharpened and properly set. It really came about that we got a distant job and had to get lodgings near our work. I was rather shy at sharing a bed with one of my mates, and when we got into the bedroom he sat on the bed and said, "I'll have this side of the bed." "You have which side you like.". I did not sleep much that night and when it was time to get up my bedmate got his big toe under the arch of my foot and I was soon out of bed. He tried to do the same again the following morning, but my foot was tight against the bottom of the bed and he could not manage it. He was a good chap for getting out in the morning and he and I were the first in the copse every morning out of about thirteen of us. We worked hard but were happy. Our daily aim was to cut and stack four cords a day. The measurement of one cord was logs four feet long stacked two feet high and the stack sixteen feet along the ground.

I remember one nasty foggy morning we were frying our breakfast around a good fire when the cord wood carters came along. They wanted a warm around our fire. They could hardly put up with it because there was some obstruction behind them and they could not get back any further and then my old mate poured some hot fat on the fire just in front of them. This of course forced them to get out of the way quickly. As they were gipsy bred we heard a bit of language from them.

At another place in Fernhurst the cord wood carters were stuck in the mud with two horses and a cartload of wood when the owner of the land came along. The wood carters did not know he owned the land they were on, but he stuck to them and helped them to get out. They expressed their thanks to him and would have liked to give him something for helping them but said they had nothing on them. After a while they said they had a brace of the old boy's rabbits and would he like them? Of course he refused them and told them who he was and said he would see that they had a brace each week that they were working there.

While working once at Tilgate near Rusper we saw some bees going into a hollow tree that we had felled earlier that year. There was some brushwood hanging over the tree where the bees were going in so we got a chap nicknamed Dandy to cut and clear this away so that we could trim the tree up for carting away with the rest of the trees. Not knowing the bees were there he started cutting this wood clear off the tree. As he got nearer the bees' entrance they got thicker and thicker, when all of a sudden he noticed all those bees and out he came, and I think my mate and myself got a good swearing at. But it was all in good faith and he did not get stung. I later cut a hole into this tree and got the bees into a box and brought them home.

George's aptitude at timber cutting, as he called it, stood him in good stead in later years. He became justly renowned for his demonstrations of splitting open hollow trees with wild colonies ensconced in them, cutting out the combs and binding them into empty frames ready for removal in the evening and transfer to a permanent place in his home apiary. This operation demonstrated the greatest skill of beekeeping and was admired by even the most sophisticated experts. For example George performed his star turn at the Berkshire Beekeepers' Centenary Field Day in 1974 to an admiring audience, ably assisted by his old beekeeping friend, Tom Haffenden This was done in the prestigious company of George Hawthorne, Cecil Tonsley, W.E.J.Hooper, Harrison Ashforth, Arthur Dines, Oliver Field, Karl Showler, and Owen Meyer to name but a few giants of the period.

At Lodge Hill, a West Sussex Centre for Adult Education, George once split open a tree trunk at a demonstration for Australian beekeepers, led by John Guilfoyle. He picked up the queen to show it round and it flew off from the palm of his hand. Subsequently it landed on the then Director of Education for West Sussex, Dr Reed. He had only one arm and the queen began a progress up his empty sleeve. George rolled it up and retrieved the queen to universal applause. A BBC team with Frank Hennig attended this occasion.

Peter Wakeford wrote of this: 'It often happens that workmen come across a colony of bees in a tree that needs to be cut down…once you see the bees coming out you have to stop. It would be a foolhardy man indeed who'd carry on cutting. Almost invariably it would be George Wakeford, the bee-master, who'd be called out to deal with the problem. I'd go with him to do the cutting and help with the bees. One time the root of a cherry tree was affecting sewerage pipes. The tree had to come out. The work had been halted when the first bees began to appear from the trunk. George had been notified so I took my car which was bigger than his and we went off to see. It was George's custom to wait until evening, the air had cooled somewhat and the bees were settled, then to block the entrance with grass. George would say 'We must judge how far the bees are in the trunk', i.e. either way from the outlet, so he'd run his ear along the trunk to give an idea of the extent of the colony. Then I'd cut through with the chain-saw to remove the section with the bees, load this section on to the car and take it away. Another day we'd split the tree with an axe and George would find the queen, get some empty hive frames, lay the frames on the ground, take the combs from the tree one by one, lay each comb on top of a frame, cut round with a knife to fit the frame, then tie two pieces of string round the frame to hold it in place. Finally he'd put each frame into the hive, put the queen with the combs, then shut the hive and let the rest of the flying bees make their way back into the hive.

An elm at Gunter's Bridge on a steep bank had to be cut down close to the ground. The bees started to come out of the saw-cut until everything was black in front of us like a cloud, quite frightening really. I cut an ash at Dunsfold beside the road. The top had been cut already so there was no weight to make it fall properly. We would have to put a rope round the top to pull it the way we wanted it to fall. We had no ladder with us so George pulled up his Mini hard by the ash, and in a trice was standing on the car roof attaching the rope to the stump of the tree. I could see the convex roof of the car buckling under his weight!

We often dealt with swarms in old buildings which were delaying the builders. When there was a swarm in a chimney stack George would have a cloth soaked in carbolic or some similar substance. He would attach the cloth to a rod and poke it up the chimney – or if he wanted to keep the bees (as he often did) he'd get up on to the chimney and poke the cloth down so that they would come out on to another cloth laid out on the hearth, then up an incline into a skep.

Stripping the bark from an oak for tanning

We cut some very large oaks at Holbrook near Horsham and as it was the spring time of the year we prised all the bark off the trees and branches as this is used for tanning leather. [According to Gilbert White only possible when the sap is rising.] After getting the bark off the trees we then stood it up to get dry, and when dry it is tied into large bundles. The man who does this is the bark tier. I don't remember this one's name. He was a very busy man and really rushed into his work. One morning we were off along the ride to our work when we came across his sack with his ties and a hand- bill in it, and thought let's have a game with this. So we tied the sack to a hazel stem so that he couldn't see the tie. He grabbed the sack as he walked, only to have it pulled out of his hand. He grabbed it again only to be served the same. Later on that day we were having lunch when he came along and we told him that Dandy was the one who always got into mischief, but the truth was that Dandy got the blame and he was the only one who did not know anything about it.

I continued to work with the trees and took fair care with my money so when I wanted to marry I was able to buy a piece of land. I went to London where there were about four bungalows built and we chose one that we liked. I paid a deposit and they gave me a plan of the footings. The bungalow was subsequently delivered at the chosen site in Wisborough Green and I helped to erect it.

George paid £55 for an extensive strip of land and £205 for the structure. The bungalow, the hall of which gave access to all the rooms, was displayed at the Ideal Home Exhibition.

Aerial view of George's home at Wisborough Green with hives at the rear

Chapter III

MY BEGINNING WITH BEES

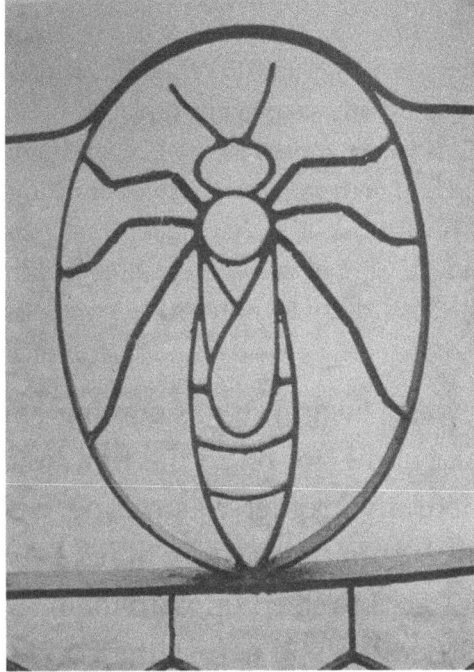

My first lot of bees was in a skip, a real old skep, and the price I paid was one golden half-sovereign. That was the recognised price in those days. If you didn't give that amount and you didn't pay in gold, they reckoned they would not live. That was somewhere about 1914 or 1915. This lot of bees died in the spring with what they then called Isle of Wight disease, now known as Acarine disease.

They didn't really know what the cause of it was then, and there was just no cure for it. We know now that it is a mite which gets in the bees' breathing tubes and makes them too weak to fly. But there is a cure for it now, some mixture of petrol in it, but I don't know the rest. You can buy it for your bees and it's called the Frow mixture. You put it on a pad, then over the bees with a piece of glass on top. The vapour goes down through the bees, it doesn't harm the bees but kills any mites which might be in the bees' breathing tubes. This pad wants to be on the bees eight or nine days and then you take it off. It is rather objectionable to the bees, as it has quite a strong smell, but it does get them free of the mites and then they carry on all right.

I of W Disease raged between 1905 and 1919 and destroyed large numbers of the native British black bees, which were often replaced by more resistant strains, mainly from Holland. Nosema spores in the bee's gut and acarine mites were blamed for the losses. Though they may well have contributed to the devastation, the primary cause is now thought to have been malignant viruses. The Frow mixture is no longer used. George was spared dealing with the depredations of the varroa mite which came to Britain only in 1992 with similar losses to the I of W plague.

The next lot I got was a swarm. I was coming back from Petworth once on a push bike and I got on to the first field near the road to the farmhouse where my brother was rolling with two horses and we looked up and there comes a swarm of bees straight out, coming so fast we were not able to keep up with them. I went through the hedge as quick as I could, and across the corner of that meadow, a meadow belonging to Mr. Johnson of Redhill House, Petworth. They went clean over there and then over the common and I had to scramble through where there were no roads or anything but I kept them in sight. When they got to our neighbour's ground they hovered round the chimneys for some little time, but on they went again right across the field, a field of peas, I remember, and there was some nice dusty ground there and I grabbed handfuls of this and threw into the air in front of the bees, but the bees just come flying right through it and on again.

The farmer, Mr. John Baker of Osiers, saw me going across the ground. He didn't recognise me in the distance. but he could see whoever it was throwing this earth into the air and he must have thought someone had gone scatty, and I suppose well enough he might. At any rate it didn't stop them at all.

George, in his later experienced years, was obviously amused to recall his youthful trust in the old skep beekeepers' belief that, because bees do not fly in rain, a spray will cause them to settle and cluster together. If no water was to hand they believed that handfuls of dirt or dust would do the trick equally well.

I didn't expect it would immediately and on they went across and over the hedge and I had to scramble through. They were making for Pheasant Copse in Lord Leconfield's Park, but I got on the road in front of them and then they seemed to hang back in the bushes before they got to the road and they settled there. What I had to do then was to walk well over a mile home and collect a skep and a sack to put the skep on and I expect I brought my smoker and that back with me and shook those bees into the skep. I then left them in the shade to settle down and I went back in the evening to tie the sack round the mouth of the skep and brought them home and set it up on legs as my father used to do.

They were such a kindly lot of bees that I left them in the skep for two or three years and I used to get three or four swarms from them each year. Now during this time I bought live or six hives and put the swarms into them. My Dad was not at all pleased with me getting box hives, as he called them, especially when they were fifty shillings each. He used to call them 'box hives' as I said and that was the name given to these hives by a good many of the older people in those times before they really got used to them and then called them bar-frame hives.

[George Senior had a point. £2 - 10s was a whole week's wages for a farm worker.]

At about the third year of my bee-keeping I had thirteen or fourteen colonies of bees each autumn, and owing to the Isle of Wight disease or Acarine, as we would call it now, I would lose six or seven of those colonies every year with the same disease. I rather think it might have been a good thing I did lose them because I didn't have any control on the number I was keeping. At the time I couldn't find the Queens and each swarm was an extra. I used to read books about the bees. I had gloves and a smoker but I was not able to examine the bees. I used to go to the hive and open up, using the smoker as I thought proper but I couldn't have been using it properly for when they started to come out I tried to knock them in again. Instead of only one or two being out there were a dozen or more flying around and I thought it was time I closed them down and got off and left them alone.

I somehow met a bee-keeper from Petworth and he came in and showed me how to smoke the bees and get them docile, and to find the queen. After that I could examine

George examines a hive, with the smoker between his thighs

them and usually find the Queens. I have quite an eye for anything that's different or unusual and also to even find four leaved clovers. I had my first extractor in 1921 as a twenty-first birthday present from my mother. This was a chain- driven affair that made quite a lot of noise and my brothers called it "the tractor" owing to the noise. This was the year I joined the Sussex Bee-Keeping Association and I've been a member ever since

Inside view of a honey extractor

George earned a modest living by helping others to look after their colonies and often managing their hives completely. But he was such an enthusiast for the craft that he loved to counsel others, particularly young beginners, offering generous help to those who asked for assistance. He may not have been the most expert of instructors but he took pains to explain and showed endless patience with anxious and inept youngsters. He was equally eager to introduce adults to the craft. As an example he set up Miss 'Posy' Heath, of Ebernoe House, with her first hive which he carried there a mile and a half on his back. She became a keen convert to beekeeping and for 24 years was Chairman of the Wisborough Green BKA, one of the five such local divisions in West Sussex. Admiral Sir Herbert and Lady Heath took over Ebernoe House from the Elins in 1923 and Posy was their one daughter.

Encouraging a young beekeeper

When George was 30 he was married on 21st April, 1930 to Maggie Sopp, the Leconfield Estate Head gamekeeper's daughter and four years his senior and who outlived him. Maggie is remembered as a good mother to their three children, David, Cyril and Josie, but also as a strict uncompromising lady, somewhat given to scold George with what was known as 'a slice of tongue pie'. He may well have taken refuge out of the house in his long hours visiting apiaries on farms, rectories and villagers' gardens for many miles around. Maggie's temper may equally been sorely tried by George's lengthy absences, but his food was always on the table at fixed times, and he was expected to be there to eat it. Reversing the usual order of things Maggie would put the children to bed and dig the vegetable garden till it was dark so that George was relieved of the labour and could get on with the planting. By tradition bee-men avoided growing the herb Rosemary as it was said to be evidence that the wife 'wore the breeches'. George omits the herb from his notes on plants, and writes little of womankind, though he was always gallant and helpful to lady beekeepers.

Here the current President of the Wisborough Green local beekeeping association, Roger Patterson, recalls his personal debt to his late mentor:

I probably know more about George Wakeford's beekeeping than anyone else and am delighted to recall some of my experiences here. In my early years in beekeeping I spent most of my spare time in his company and he has been a great influence on me. Because of his charismatic personality and his general knowledge of country life he was well respected by everyone who saw him at work. I was brought up on a farm and knew hard times just as George did, so I understand his approach to things. If you need something you improvise before buying; if it is broken, you mend it rather than throw it away and buy another.

I started beekeeping in June 1963 [when George was 63] soon after the 1962/3 winter when we had snow on the ground for 14 weeks. I had found a colony of bees on a neighbouring farm but I didn't at that time have much knowledge of beekeeping. My first biology lesson at the Weald Secondary Modern School at Billingshurst at the age of 11 was on honeybees. Being used to wildlife all around me I remembered the lesson, so when I discovered the bees four years later I enquired about them and bought them. I contacted the biology master Allan Dugdale for advice. We looked at the bees and, although they had some honey, he considered they were too aggressive for a beginner so he changed the queen to give me a more docile colony. Allan advised me to join the Wisborough Green Association which I did there and then. We took the honey off the colony but I would need to get the honey jars from the Secretary who was George Wakeford which is how I came to meet him for the first time. He was most apologetic, 'I'm dreadfully sorry, I can't let you have any as they are only for members'. He didn't know I was already a member. When I cleared this up I got my jars.

A couple of months later, when I was putting some of my newly acquired honey into the Billingshurst Flower Show, George was exhibiting too but there was no question of rivalry. He carefully showed me where I was going wrong, putting as much right on the day as he could. The trouble he took with my exhibits you'd hardly think he had exhibits of his own. I didn't win a prize but I learnt a great deal.

I realised there was an opportunity, so asked him if I could help him and he readily agreed. This gave me the chance to learn, but George was not a good instructor because he was quiet and assumed people knew more than they did and would pick things up by observation. He gave me very little tuition but having a practical background I learnt a tremendous amount simply by watching him. I later realised that George really did not know very much about what was happening inside a beehive. Very few beekeepers did at that time as there weren't the pests and diseases the modern beekeeper has to deal with on a constant basis. George knew how to recognise notifiable diseases and how to deal with them, but not how they were caused and how they worked. He never read a bee-book, learning by observation and his contacts with other beekeepers. In those days that was enough to manage bees, get honey and keep them alive, which he did very successfully. There wasn't the time to study, as beekeepers have to do today. As soon as he got home there were things to do, either feed and water the chickens or do some gardening. George's strength was he was the best handler of bees I have ever seen, with nobody coming anywhere near his ability. It was a joy to watch him and see the way he controlled a colony with a few puffs of his smoker. He "read" a colony brilliantly and responded instantly to their needs. This is stockmanship at its best and it didn't matter if it was cattle, sheep, chicken or bees, he was good at them all.

I have been teaching beekeeping for a long time and believe the techniques of beekeepers are formed at an early stage. I largely use the methods I learned from George, but have modified or improved them to suit changing circumstances. I use those methods when teaching others and think the Wisborough Green members are very good compared to others I see; they are my methods developed and refined from George Wakeford's.

George was a founder member of the Wisborough Green Division of the Sussex Beekeepers' Association in 1947 and was Secretary for about 20 years from the early 1950s. The Chairman during most of this time was Miss Heath of Ebernoe House who George had helped start beekeeping by carrying a hive on his back from Palfrey Farm. The situation was not ideal because George had little secretarial ability and nobody was strong enough to relieve him of the post and make more use of his considerable beekeeping skills in other ways. There is no record of a committee meeting between 1957 and 1963! If anything needed doing George just got on and did it. The Treasurer at that time was another well-known beekeeper, Tom Haffenden, whose administrative skills were similar to George's. He would present all he had to the auditor and if things did not add up he either took money out of his back pocket or put it in! There was absolutely no suggestion of dishonesty, but a reflection of how unaccountably things were conducted in those days. Considering most transactions were made by cash it was so easy to put a subscription or money for honey jars into your back pocket and forget about it.

George looked after bees for other people for the main part of his income. His clients varied considerably from the person who had only one colony to those with 20 or 30 for pollinating on some of the fruit farms then in existence. These people came from all walks of life and many colonies were owned by people of limited means. I didn't get involved much in his finances, but I detected that, considering his skill, he was probably not charging anywhere near enough. His attitude to matters of finance was relaxed. He'd send in a bill for his services and that was that. Nothing would ever induce him to send another one. He once let it slip that although he had sent in bills someone hadn't paid him for three years. I knew how hard he worked and although it was none of my business I showed my concern. His reply was, 'The amount of money they deals with, boy, makes mine small and they don't bother about it!' Similarly if he lent equipment he would never ask for its return. As he would say about his bills, 'Well, if they wants it that bad, let them have it!' Some of the problem was his failing memory. In those days everyone accepted it whereas now people seek help. Perhaps his condition could have been managed so that he did not suffer so much in his last few years.

His relationship with the colonies he managed varied from person to person. For some he simply made a regular call at the beginning of the season to put on the 'supers' or honey chambers. He would also call if a person had a swarm and then extract the honey at the end of the season. Others, such as fruit farms, needed regular attention. He never kept a diary, and as his memory was failing there were some chaotic incidents, though often amusing.

Extracting the honey was a familiar job, as was feeding the bees for the winter. He did the honey extracting in people's homes, taking his own equipment, in houses or other buildings such as garages, greenhouses, farm buildings and in one case a large chicken shed. The levels of hygiene did not match what we have to comply with today and I have seen a lot of honey scraped off dirt floors. The usual comment was, 'It's only clean dirt, boy.'

In the mid 1960s I reckoned George had charge of over 400 colonies. His main summer work was dealing with swarms and the pattern of each day was similar. His wife took the 'phone calls and in the morning he would deal with the previous afternoon calls. Then he would come home for 1 p.m. lunch, and in the afternoon deal with the morning calls. There was little pattern and I've known him go as far as Shermanbury and Henfield to Wormley in Surrey, a distance of some thirty miles in the same day. A great saying of his was, 'Got to get they wheels turning, boy', but with little attention to making efficient use of time. He was of an age when driving tests weren't compulsory and sitting in the passenger seat of his Ford Prefect (KVJ 281) was an experience that can best be described as character-building. He continually slipped the clutch and, although his car barely did 60 mph flat out, he would hit the floor with his foot to get up reasonable speed, then 'coast' for a distance, then repeat it. This practice was commonplace in those days but didn't save the fuel it was thought to. He was rather inclined to let his attention wander to the roadside. 'Look at that...', he'd say, pointing to some plant or bird without taking into consideration anything else that might be on the road, or even if there was any road left! Nature fascinated him, but did not improve his driving!

Because George was such a good handler of bees his ability allowed him to control colonies that others would find difficult or impossible. This led him to keep some colonies that the modern beekeeper would be well advised to requeen in order to make them more docile. I saw him give up on bees on only one occasion because of their bad temper. He kept six or eight colonies permanently at High Noons fruit farm at Balls Cross. They were backed up against a hedge in a field with bullocks in, enclosed in a barbed wire fence to protect them. We were passing by and we got over the fence to inspect them, beginning at the furthest one. We found that the first colony was tetchy, the second more so and the third even worse. By the time we opened the third we were under attack from the first two as well. It could only get worse as we proceeded along the row. We were badly stung and decided that any damage done caused by negotiating the barbed-wire fence in a hurry would probably be less than the bees were inflicting. I have handled many vicious stocks since but that was certainly a learning experience! The wearing of full protection, which is now the norm, doesn't alert the beekeeper when the bees are aggressive and thick gloves makes the handler clumsy which upsets them. George wore no protection at all which concentrates the mind.

If another beekeeper sought his help with a spiteful stock George would substitute one of his own gentle queens for the vicious mother. She would go back into his home apiary so that some of his own stocks were unpleasant to work. He always reckoned that the first two or three stings of the season hurt him but after that had no effect. His gentle handling meant that he was not stung as much as other beekeepers. He never made a fuss but was always quiet and deliberate in his movements, a lesson I learned from him at a very early stage.

Working with him led me to know some of the Wisborough Green BKA members and his clients quite well. Although the latter would allow a teenager to accompany George some wouldn't let me into their houses. He didn't rely on me but I was clearly becoming useful, especially at extracting time. Not having my help then made the job much longer for him.

At that time the only beekeeping teaching at the Wisborough Green association was about three summer meetings at members' apiaries. The standard format was George inspecting the bees before tea was taken. He made few comments and no descriptions of what he was doing. He would let me help as he would at his other hives. In 1965 I was elected to the Committee, being about 40 years younger than the next youngest! I soon realized that these meetings were little more than social affairs with members watching George inspecting colonies, some of which were pretty grim. I suggested to him that perhaps we should have our own association apiary. 'What do you mean, boy?' was his response. I explained that we could have a permanent place where we could regularly see the same colonies and observe their development through the season which would help members and beginners like me.

George said little, but a few weeks later revealed that he had found a suitable site at Dounhurst Fruit Farm. He hadn't even consulted the Committee but gone ahead and done it! There were immediately offers of four colonies to stock it, though it seemed a handy way to get rid of bad-tempered bees in ramshackle hives. The association apiary has now been at Dounhurst for over 45 years and may be the largest and best in the country. It has trained many hundreds of beekeepers in mainly the same methods that George developed and used so successfully and for so long. It was named The George Wakeford Memorial Apiary.

I was elected to go with George and Tom Haffenden as a third delegate to Sussex BKA meetings, usually at Worthing. George and Tom, both unpredictable chauffeurs, took turns at driving. Their conversations were always the same though, with the same tales but in differing order. There was always room for this exchange:

George: Well mas' Haffenden, what's on the agenda tonight?

Tom: Dunno met.

George: Well, whatever they says, I shall agree with them!

As a raw teenager I struggled to see the point of putting three people's lives at risk on 40 odd miles of road, just to agree with others.

George, with one of his bee-mobiles

When Miss Heath died in 1970, after being Chairman for 24 years I was elected at a crisis meeting at George's suggestion and served in that capacity for 26 years. I got on well with George who accepted my ideas. Recognising that he would be glad of losing the burden of Secretaryship we thought it would be a proper reward for his long years of good work to ask him to fill the soon-to-be vacant role of President. His protests that he was 'not worthy' were politely ignored and a fairly new member, Sunny Warner, replaced him as Secretary. When after a couple of years he asked to step down 'to let someone else have a go' he was elected Life President, to which he had no answer.

Some 50% of the members were George's clients. Few of them knew much about it or why they received the newsletter. He just added the subscription to the bill at the end of the year. As Chairman I concentrated on building up the teaching apiary and allowing members to handle the bees rather than just watching the Bee-master. The membership increased and I started demonstrating, sharing the work with George, a format we still use 40 years later.

About 1972 Wisborough Green BKA was one of four divisions to withdraw from the Sussex BKA to form West Sussex BKA. George served one spell as President and then ceased involvement in beekeeping administration.

By the time I met him he'd long given up full-time timber cutting which he'd done for years, away from home for nights at a time, sleeping under canvas as often as not, when the weather was warm. He had a commendable attitude towards work. He was still putting in seven days a week with bees and gardening. During the winter he would also pick up several other jobs, though mainly gardening, his knowledge being often called on for such things as the pruning of fruit trees. I was involved with him in other ways. He had occasional winter work that involved cutting timber when his son David and I would help him. One regular job was to thin out oak saplings called tillers at Barkfold Manor, Kirdford and cut them up for firewood. It would have been better to split them for fencing posts and rails, but George always did what he was asked. He never used a chainsaw, preferring a crosscut, normally operated by two men, one at each end. On large trees there were two more, the outside men pulling ropes tied to the saw. He heard of a second-hand chainsaw which would help with one job and persuaded me to buy it. Though he would never use it himself we used it often to make life easier for him.

I remember pollarding a large lime at the Women's Hall in Billingshurst where the Trustees had a quote for £47.10s which they thought too expensive. George and I did it for £10. It took us both two days. Though he was in his late 60s he insisted on climbing the tree himself and cutting the branches off by hand, lowering the cut branches on the rope to me below to deal with while he cut off the next one. I had no training for such work but was familiar with such

situations so the pair of us worked well together fully confident in safety, knowing what the other was likely to do. I learnt a lot from that job which I've often used since. One winter I bought a couple of acres of standing chestnut and he worked with me making fencing stakes. It was hard work, splitting with wedges and pointing with an axe. He flatly refused to accept any payment.

George was well known as a beekeeper, so often called upon to take settled swarms or ones in cavities in trees and buildings. He had a young lad, Arthur Mann, who helped him for a little while. Once when we were taking bees out of a roof was the only time I ever heard him swear. He and I were on the flat roof taking bees out from behind a strip of lead. As we were peeling it back George asked Arthur if he would move his jacket left lying on the ground. Arthur, being a bit bored, took one corner of it and dragged it along the ground so that George's wallet came out of a pocket, spreading banknotes and cheques on the ground. If George was paid by cheque he would fold it and put it in his wallet. When he bought something like petrol he would sign the back of the cheque and use it for payment or part-payment as people did in those days.

I built my colonies up to about 130 in a dozen locations, intending to become a commercial beekeeper. George helped me a good deal and wouldn't accept any payment at all. This was typical of his generosity and I shall be forever grateful for his guidance and wisdom. I couldn't have done it without a sound grounding on handling bees and simple management techniques.

George never forgot his upbringing at Palfrey and would go back to Ebernoe Common at the slightest chance. I kept some bees there for some ten years that did exceptionally well. As the area was mature woodland there were many colonies in hollow trees, probably native type bees that were survivors, unlike imported bees that don't suit our climate well. Any swarms I got from them were usually good bees so I was keen to get as many as I could. Ebernoe Common was then part of the Leconfield Estate, but is now a nature reserve owned and managed by the Sussex Wildlife Trust. After we had inspected the hives George would show me the various places, regularly mentioning that in his boyhood there were 13 cottages on the Common, but now only one. One was Birchwells where his Grandmother had lived.

Adjacent to Ebernoe Common is a house called Little Wassell. It had a dirt floor and Mrs. Poste lived there. She had bees that George looked after. I visited several times. He always called her Mrs. Poste and she called him Mr. Wakeford. One day George said, 'Mrs. Poste and I are related in some way', so I asked how. His reply, 'Her mother and mine were sisters'. I am a bit suspicious about it since he had not recognised they were cousins. Also they would both have been born in the later 19th century when civil registrations were not always accurate or recorded. In many communities the male parentage may not have been what it appeared to the outside world!

George was always fun to be with and I rarely saw him show any anger. He thought well of everyone and when there was a problem looked for an excuse for them. Like most country folk he had tales to tell, usually poking fun at those in an influential position e.g. policeman, vicar, doctor, etc. It could become a bit repetitive, but people laughed and there was no malice in it. One time an annual beekeeping weekend was organized by West Sussex County Council at their Lodge Hill Residential Centre at Watersfield. Arthur Curnuck, the Advisor, thought it would be a good idea if George told a few tales after dinner. George's wife, who was not keen on bees, was asked to attend. She had difficulty in keeping her thoughts to herself. After dinner George was introduced. He'd got a few words out of the first tale. Those of us who knew him recognized it, including his wife who said in rather a loud voice, 'You can't tell that one, George!' His hearing was beginning to fail, so he ploughed on. His wife was even louder. 'George, George, you can't tell that one. Whatever next!' This was repeated with every tale and provided much more entertainment than we expected to get in the first place. No doubt many of his tales were recycled from elsewhere but many were genuine and happened to him or witnessed by him.

One amusing thing happened to both of us. One evening we two were to move three WBC hives from West Chiltington to Dounhurst Farm at Wisborough. WBCs are double-walled hives with an outer case and brood box and supers with the bees inside. In good flying weather the bees are normally shut in and moved in the evening when they have stopped flying. We drove to where the hives were.

There was a cutting with a 12-15 feet high bank and a drive at the bottom, possibly once the entrance to a large house. The steep bank had a hedge at the top and a flat area where the hives were. The only way to get the bees to the car was down the bank. We closed up the first, tied it up and got it down to the car. George's Ford Prefect, with the sloping back took one hive nicely. After the first of our eight mile journeys without incident it started to rain. When we got back the slope was getting a little slippery. We secured the second hive and with some effort and a little slipping got it down, into the car and off to Dounhurst. The rain got progressively harder so by the time we got back for the third it was tipping it down. We shut it up, tied it and tried to lift it but it was much heavier than the others, doubtless weighty with stored honey. We struggled to get it down the bank and slid most of the way. We struggled to get it on the car and off to Dounhurst, then strained our backs to get it into place and untied it. George looked inside to see what made it so heavy only to reveal that it was full of bricks. George was a well-loved character, and as with all characters, there was usually a bit of mischief not far away. There were many stories told about him, some true, many distorted and often by people who had never met him. One common theme is that he had a queen bee in his cap. Yes, he often did, but few knew why. As much of his summer bee work involved colonies that were swarming, there were either fertile queens spare which he would put in his cap in case he needed them for another hive later in the day or, as many colonies would have queen cells, he would put them in his cap to keep them warm. Some of the cells were so advanced the virgin queens had often emerged by the time he took his cap off again. Often he would have more than one queen alive in his cap. It always amazed me that they never fought to the death as they would have done in a hive. He would take his cap off carefully by pulling it forwards, so scraping the back of it over his head. This ensured all the queens stayed inside. I never saw one fall to the ground.

Although I knew George and his wife for over 20 years I never knew Maggie's name for sure till researching this story. George usually referred to her as 'The old Dutch' or 'The wife'. Everyone called her Mrs. Wakefield, though David, their son called her Meg occasionally. She was quite keen on buying things off hawkers. George was rather concerned about this, so instead of saying anything to her he put a colony of bees on the lawn by the path leading to the front door. This stopped it!

George's sister-in-law, Edie Tickner, lived with them for some time, though I never knew the circumstances. Once when we were in West Chiltington he said he was going to see his cousin Ethel someone unknown to me. We fought our way through the bushes to get to the back door of her bungalow and found her in a very poor state. We did what we could to help her and soon afterwards George went to collect her. He brought her back to Wisborough Green and I believe she lived with them for the rest of her life. Those two incidents show how generous George was.

George rarely used Christian names, mainly from politeness but also because his memory was failing. He would regularly refer to people as ' Mr. and Mrs. er um er'! Even Tom Haffenden, his old friend and five years older would be addressed as 'Mas' 'affenden. There were nearly fifty years between us and he would call me 'boy' as he would anyone of a younger generation. Bees he'd call 'They little people' or 'They little jiggers'. He used words that were probably relics from Sussex dialect, one of which was 'snotchell', which means a small amount.

Like me, he was a village cricketer, playing first for Ebernoe and then Wisborough Green. As he worked six days a week as a timber cutter I suspect he did not play on Saturdays. He must have played well into his fifties, stopping before I started, though I remember his boots hanging up in his shed. People I spoke to gave him the reputation of being a safe catcher with hands like buckets. The physical work he did was enough to keep him fit to play.

He was a very good gardener and always had something in flower or available to eat. His generosity was such that visitors would come away with an armful of vegetables or a box of apples. Despite his onerous workload his garden and orchard were exceptionally productive. He wasn't afraid to grow something different. I remember him growing mercury and salsify, neither of which is very popular. He kept his bees in ramshackle hives in his orchard together with several chickens and an occasional goose. The apples were productive and well pruned and included several uncommon varieties such as Ecklinville Seedling, Adams Pearmain, Gladstone, Royal Jubilee, Forge, Lady Sudeley and one he had bought as Blenheim Orange, although it clearly wasn't. For a time he was convinced it was Winter Quarrenden, but it wasn't that either. It resembles a Cornish apple called Ben's Red, and despite my search for information I'm not convinced of that either. The one George had could readily be struck from cuttings and bore fruit at an early age. He pushed the prunings into the edge of his garden every year and distributed many young trees, one of which I still have. It is a good cropper of September/October dessert apples, but is susceptible to bitter pit and is a tip bearer rather than the more common spur bearer, a fact not mentioned elsewhere, hence my suspicion.

With the mechanization and computerization of what little agriculture there is left, sadly we shall never see characters like George Wakeford again. He could turn his hand to anything including ditch digging, coppicing and hedge laying. Modern culture, with the throwaway society, globalization, reliance on electronics, health and safety issues, blame, legislation, litigation, selfishness etc. have created people who are largely impractical and not allowed or expected to think for themselves, or develop such abilities as lateral thinking. It's called progress but it inhibits the natural abilities of people to develop the skills that made characters such as George. I consider myself incredibly lucky to have known him. He taught me so much, not just about bees but about many other things, including how to treat people well. I will always remember him laughing. It is not by accident my two sons are called George and Tom.

A handful of bees

Roger's memoir raises an interesting issue of the traits that qualify a person as a 'good teacher'. One must accept Roger's view that George lacked instructional and demonstrating abilities, yet in almost every paragraph he records how much he learnt from him. There is an apparent contradiction here only resolvable by realising that good teaching can be achieved without verbal explanations such as might delight a visiting Inspector of Schools. Dickens reminds us by his portrait of Mr.Gradgrind that true learning is not the mere cramming in of facts, the conning by rote that a horse is a 'Quadruped, Graminivorous. Forty teeth etc.' and teaching is not the filling up of 'empty vessels', but the development of the imagination and a sense of wonder at whatever aspect of our magical universe we may choose to explore. Most of our early learning is imitative. We learn by copying what we see before we acquire language and we acquire words and body language by imitation too.

Mark Pagel, Professor of Biological Science, argues that all human culture has accrued, ever since the earliest homo sapiens, from cooperation within social groups where a few innovators have been followed by the vast majority of imitators. Our distinguishing human intelligence is our ability to assess what others are doing and copy the best examples. Culture is an imitation system where nothing matters more than reputation. George enjoyed such a reputation and apiculture is but a tiny twiglet in the vast panoply of modern general culture. When we add together the capacity to show a worthy example to a pupil, to engage the beginner's friendly trust and confidence and the gift of inspiring enthusiasm and a sense of wonder, then we have a formula for a great teacher, however deficient that person may be in purely instructional skills. For all its faults, learning from observation may have more permanent value to a pupil than factual knowledge mugged up for an exam, especially when the person observed is liked, respected, competent and worthy of trust. Nothing is more valuable for true learning than the urge to find out and understand for oneself.

If a client asked for help with a spiteful stock George would substitute one of his own gentle queens for the vicious mother of the offending colony, so curing the problem. The sharp breeder would go back into his home apiary, so as a result his own stocks had a reputation for stinginess. There is a belief among some beekeepers that wicked stocks make the best honey gatherers, so George may have gained a little benefit from his own generosity.

His formal mode of address was no doubt due to his role as a specialist who made 'home visits' similar to that of a doctor, district nurse or vetinary surgeon. As an all-wise crutch in emergencies to anxious novices he was accorded the deference owed to his expertise. There was much scope for necessary intimate dialogue and trust unusual in the experience of a rural working man. The fancy for producing honey transcended all the prevalent hide-bound class distinctions. Consequently George was obliged to socialize daily with the aristocracy, squirarchy, clergy, professionals, yeoman farmers, tradesmen and village cottagers alike. He treated them all equally with the same appropriate courtesy and respect.

The Squire, John Peachey, and his coachman Mr. David Fiest

George's indifference to the profit motive, or disregard of the main chance is exemplified in the ludicrous rates he charged for his own services. It is widely believed that he casually handed over valuable property, jewels and golden sovereigns he had taken care of and expected to inherit from an old man whom he alone had cared for at home and in hospital. And when he did receive a small legacy, he at once divided it between his children, keeping only enough for himself to buy a colour television set, then newly in vogue.

The Lore of Bees

It used to be said that if the bees were not paid for in gold they would not live, so the bees and the skep had to be paid for with a golden half-sovereign.

If the owner died the bees were told in these words "Bees, bees, bees, your master is dead. Will you work for me? Then the hive was tapped three times and a piece of crepe was tied on the skep.

Superstitious folk also believed that bees were doomed if they settled on dead wood or were carried over running water! The bees had to be told of a change of ownership, marriage or any other family event.

Marriage, birth or buryin,
News across the seas,
All your sad or merryin,
You must tell the bees

Rudyard Kipling

When a swarm came out a fire shovel was beaten with a key. This was a way of ringing the bees swarm to get it to settle. If the swarm should move off they would be followed and the ringing continued so that if the swarm was seen by another person he would not claim them because he would have heard that the owner was in close pursuit.

Tanging, or the 'tin can music adopted on such occasions' accorded with one of the laws of King Alfred the Great. 'The swarm is the property of the original owner as long as he can keep it in sight; afterwards it becomes the property of the first person who secures it'. Plato, Pliny and Virgil all believed that swarms could be settled by 'the beating of brazen implements', but George knew better.

I have done all this in the past, but not again because I feel sure there is nothing in it. If the ringing is done really close to the bees it might drown the noise made by the Queen's wings, but what part of the swarm does the Queen occupy and how do the bees know where the Queen is as they fly along? She always emits a scent and

I feel that this scent is much stronger at swarming time than at any other time.

George was referring to the pheromone or chemical scent called 'queen substance' (or 902), a complex acid produced in the queen's mandibular gland, which controls the behaviour of the colony as regards swarming and mating. This was first described by Dr. D.C. Butler just two years before George wrote his book and shed a remarkable new light on aspects of bee behaviour hitherto a mystery to beekeepers. George knew little about what prompted the bees' activities inside a colony. His knowledge of bee behaviour was gained by astute observation and practical experience.

I have found that a queen from a swarm is more readily accepted by queenless bees.

It is generally understood that you can go on anyone's property to get your swarm with permission, but that if you do any damage in getting your bees you are held responsible for it. If you lose sight of your swarm, anyone finding them can claim them.

George taking a swarm from under a currant bush

Running a swarm into a hive

When a swarm leaves a hive they usually settle in a ball, and then is the time to shake them into a skep. If you do not take them they will usually stay there until about 11 a.m. the next day. Some of these worker bees are very young, so where better for them to mature and get ready for a flight of about two miles than in the middle of a ball of bees? They most always fly towards the sun. Perhaps they do this to find a veritable land of milk and honey. When the weather is fine and warm and nectar is plentiful, they work like demons to gather all they can. I think they know that if they do not store enough to last the winter they will starve. I also think that they would show up well with the seven good years and seven bad ones. If their honey was not taken from them during seven good years it would keep better than any grain grown or even perhaps wines and spirits. I think good honey would do anyone more good.

George said that he had a spoonful of honey each day and reckoned it great for keeping him feeling young, fit and 'regular'.

Surprisingly George nowhere writes of his skill as a skepmaker. He often demonstrated this ability at beekeepers' meetings, encouraging others to take up the old Anglo-Saxon craft. No doubt there were many more sophisticated practitioners who made skeps for profit, but George was a competent expert. In his youth he will have learnt to make dome-shaped skeps, since they were the only permanent hives cottage beekeepers knew of. They could be made at little expense, long-stemmed wheat straw being freely available on arable farms up to the 1960's. Stripped split bramble stems provided the binding strings. The simple tools needed were just two. A 'feeder' would be made from a piece of hollow bone or a polished ring cut from a cow's horn which was used for progressing strands of straw to create a coiled roll. A large eyed bodkin was used to sew the coils together to shape up the skep.

Later in his life, when skep hives had been largely abandoned in favour of WBC's and Nationals in England, beekeepers still had to collect swarms, however hard they tried to prevent them. Light skeps, usually with flat tops, were still much in demand for this purpose. Their slightly flexible walls allowed them to be squeezed into awkward spaces under swarms. They were light enough to be handy up ladders and could be pronged on a pitchfork to be hoisted aloft under a cluster on a branch. When the queen and most of the bees were taken the skep would be up-ended on a sheet, so as to collect the flying stragglers, ready to be gathered up at dusk for hiving back at the home apiary. They would be comfortable in the dark cavity which, when well-used, would have scents and bits of comb encouraging them not to decamp. More importantly, unlike buckets and plastic bins, the porous walls allowed hot air to escape and fresh air to circulate, so avoiding mass slaughter from suffocation.

George making straw skeps

My Garden

I like my garden. The front has many including roses and clematis. As I write this on February 26th, 1976, my Mahonia "Charity" is a mass of blossom scenting the whole garden with a lily-of-the-valley smell, and it is of course a favourite of the bees. Witch Hazel is making a good show and the Veronica has been in flower all the winter. Snow- drops and crocus are in full flower and provide food for the bees. Helleborus Niger is well out and well visited by the bees. Iris Stylosa has given more flowers this year than in the ten previous years put together, with picking of twenty at a time from one plant. Datura Cornigera, "Angels's Trumpet" in a heated greenhouse, safe from frost, is in full flower. I saw the first single willow catkin a week ago and others are coming into flower today at Thakeham, while the first coltsfoot was out a week ago, both good for the bees.

In the vegetable garden at the back there is no weed to be seen-not yet. Broad beans and peas are well up. Turnips from last year's sowings are still available, and there are cabbage, sprouts, sprout tops and broccoli to follow. I grow three sorts of French beans but the blue vine Cocoa bean is best for the deep freeze. I grow Jerusalem Artichokes, and a plant called Mercury. This is a spinach flavoured vegetable which lives for years, is very hardy and will grow among grass and stinging nettles. You can pick the tender leaves and stems to eat when there is very little else about. [Goosefoot Mercury is better known as Good King Henry]

I will now let you know how to get rid of five of our worst weeds, namely land thistles and creeping ones, sow thistle seed, creeping rock and milk lilies or convolvulus and ground elder, in that order. Land thistles can be completely eradicated with a sharp hoe. Every time you see one hoe it off. They breathe through their leaves so that will kill them. Sow thistles are not deep rooted, unless you have dug them in deeply, so dig them all out and watch out for any missed pieces and seedlings. Rock lilies are harder to get rid of than the milk lily, so when they grow about six inches long spray them with Verdone, [a proprietary selective weed killer] made up exactly to the strength named on the bottle so as to kill them slowly. They take the Verdone down with them. Should any come up again, you must spray them again and maybe once more. You must also watch for seedlings, for a year or more.

Milk or greater lilies [bindweed] only grow deeply in light land and can be dug out where they can be got at; if you stick to this you will absolutely kill them all. The runners of the milk lily climb up any support that comes its way and come out into a white trumpet flower which produces seed. The runners that creep along the ground will run along to about twenty feet and then branch into about ten or twelve pieces that bore into the ground and produce another batch of young plants at this distance from the parent plant. Thus if another runner goes this distance in another direction one has to root them out carefully.

Newly dug bindweed roots

Another awful creeping weed is mare's tail. This goes very deeply into the ground and I don't know if there is a cure for it. I think perhaps if it was hoed off daily as it appeared this will finally kill it. For weeds in paths or pavings I use boiling water from the vegetable saucepan. Again one must watch for seedlings of milk lilies and mare's tail which will stay alive in the ground for some years. It seems they germinate when they are manured to about twelve inches in the ground. If they are deeper than this they stay dormant for a few more years.

You should dig out every available piece of ground elder and watch regularly for any pieces you might have missed. Dig out completely any plants it has grown into if it comes from a neighbours; it's yours when it gets on your ground. Spray it as it comes to your boundary and watch for seedlings. but all these nasty weed flowers are good for our little bees. Another but lesser weed is the coltsfoot, a weed good for early forage for the bees, but if you don't like it in your vegetable patch every piece should be dug out and a watch kept for any pieces you might have missed. All lesser weeds should be hoed off with a sharp hoe regularly, preferably in the early morning and when they are young. The hoeing in the early morning gives these weeds a full day to wither and die, especially if we get a good sunny day. If the day has rain the weeds may re-root and the hoeing will have to be done again.

Roger Patterson commented: 'George had considerable botanical knowledge, being in this, as with his bees, largely self-taught. He knew the botanical names of most wild plants without going into the technicalities of things like mosses and liverworts. He was particularly knowledgeable about apples. He was an excellent gardener, always capable, if some crop failed to come up, of producing something to make up the shortfall. There was always some flower in his garden even in the depths of winter and he knew how to get mistletoe to 'take'. But it was as a handler of bees that he will be best remembered. Perhaps he weren't a scientific beekeeper in the modern sense, but he was and will always remain the best handler of bees I have ever seen!'

One of George's West Sussex contemporaries was Fred Streeter, born not far from himself at Pulborough and Head Gardener at Petworth House who became a 'national treasure' as a radio gardening expert. Both men shared a love of plants and George's life, style language and expertise were not dissimilar from that of his more celebrated co-eval. George once claimed that Fred, who hated a challenge to his own opinions, refused to be convinced, like William Shakespeare, that the queen bee was a female and not the King. Both these remarkable Sussex men lived their lives under the umbrella of the old world Leconfield Estate centered on Petworth House. Fred became famous for his radio broadcasts. Since gardening enjoyed a much larger audience, numbered in millions, compared with beekeeping, George's escapades with media lacked the same popular appeal, though the mystery of the subject and his tricks with bees brought him to TV studios to appear on *Blue Peter* and *Nationwide*.

What endeared him to viewers was his custom of working his stocks without protective clothing and the fact that he carried queen bees about in the warmth of his trade-mark cloth cap. Even more astonishing was the story, attested by some but denied by others, that he had a trick of popping a queen into his mouth for short term safe-keeping while working a colony. It is a difficult feat to hold a heavy frame teeming with bees with one hand, isolate the queen and pick her out with the other, and open a matchbox to secure her from escape at the same time. The use of ever-ready lips, for a brave man without a veil, would be a sensible solution. Fortunately queen bees do not sting humans, though some claim they bite. A little spittle would deter her escape when she was retrieved and boxed as soon as the frames were back in the hive. It seems more likely that he occasionally held a queen in his lips by the head and thorax with the sharp end outside. Another better authenticated habit of his was to keep a few mature queen cells that he had cut out from swarming colonies in his cap. Then when occasion demanded a new queen to save a colony, found 'queenless' and so doomed, he had the remedy instantly available.

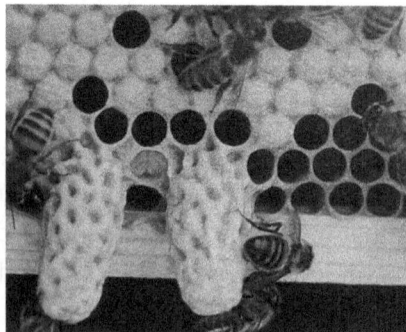

Queen cells such as George would store in his cap

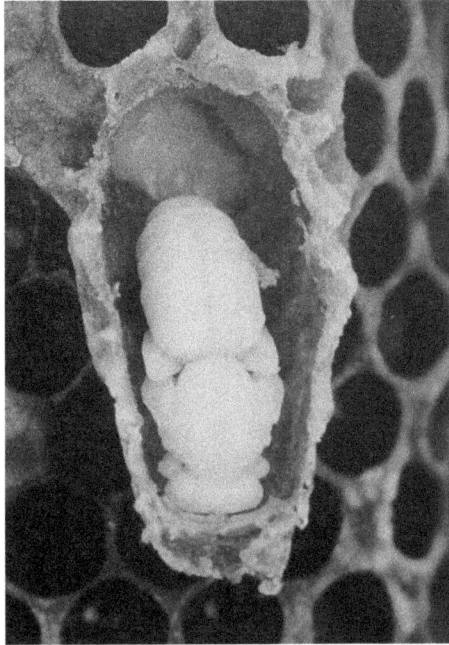

Queen cell exposed to show the immature queen larva and the royal jelly she consumes

When a worker bee implants its sting and flies away the bee's entrails are pulled out resulting in death because the sting barb is ridged rather than smooth.

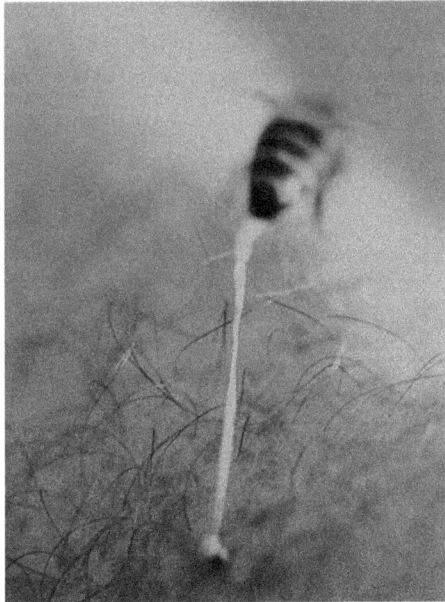

Doomed worker with its sting planted on a hand.

The worker bee's sting

George would patiently encourage such a bee squatting on his skin to twist itself round, so releasing the barb and allowing its survival.

Fred was honoured by the Queen with the award of the MBE and George the BEM presented by the Lord Lieutenant. These honours are awarded for "service in and to the community…which is outstanding in its field; or very local 'hands on' service which stands out as an example to others. In both cases, awards illuminate areas of dedicated service". As a relatively unsung local hero George was worthy of his honour and among those we should celebrate.

George with Lavinia, Duchess of Norfolk and daughter Josie

Chapter IV

TALES OF WEST SUSSEX

Rev E.M.Sidebotham conducting a Rogation-tide Service at Kirdford,1939

George's autobiography affords ample illustrations of his gentle sense of humour and love of a tale well told. In the years between the world wars wherever farm workers and craftsmen worked in teams, mowing, hoeing and performing other manual tasks, or in the brief minutes of relaxation at the start of the day when horses and jobs were allocated by the foreman, or at 'elevenses' and 'fourses', there were golden opportunities for the spinning of yarns and the reviewing of old jokes passed from generation to generation. Typically George's stories reflected the simple countryman's delight in 'putting down' the self-esteem of the pretentious and high and mighty who ruled their lives. His anecdotes most often deflate the pomp of parsons, doctors, schoolmasters, Home Guard officers, bossy hospital sisters, all the superior people regarded by villagers as their betters. These were all fair game for the subversive wit of clever but largely unschooled men, such as George Wakeford, aptly called 'village-Hampden's and 'mute, inglorious Milton's, men celebrated in local circles, but largely unrecognized outside their neighbourhood. That said, he was not above a chuckle at the expense of the irrational and confused butt such as the man who meant to look up a chimney with his eyes shut.

I would now like to go on to tell some amusing stories and local sayings of the West Sussex area in which I have lived for so many years and which are all actually true.

I will start with a story about the Reverend Sidebotham, at one time Vicar of Kirdford. It is a true story of what happened. While he was there, and he was a very nice chap too, a portion of the churchyard wall fell down and this meant raising money to put it up again. He went to a good many people, and to one in particular who was a farmer and whose land adjoined the churchyard, to ask for money to help rebuild the wall. This old chap wasn't very keen on paying up. He thought a while and then he said. "It's like this, vicar," he says, "all them that are in there can't get out and all them that are outside don't want to go in. If I was you I wouldn't bother."

I have another tale concerning the same Vicar of Kirdford.

There were four brothers of the name of Matthews and one of them died and when it was time for the funeral, Vicar Sidebotham was going to the church and one of the Matthews chaps had just come home from his work. He had his old dinner basket on his shoulder, and the Vicar said, "Why, not coming to your brother's funeral, Matthews? '' "No '' he said, "I didn't think I would." "Why ever not?" the Vicar said to him. "Well, I don't suppose he'll be coming to mine."

A tale about the Reverend Newberry of Kirdford is my next. He was going to the station to catch a train and he hadn't left himself quite enough time to get there so he was in a great hurry. When he got up to the green instead of taking the road to Billingshurst he took the road to New Pound. When he got to New Pound he still took another road coming directly away from the station. All of a sudden he realised it and turned completely round in the road in front of an articulated lorry. The lorry driver, to avoid a head-on collision, crashed into three stationary cars, and then to finish up with the reverend gentleman almost ran into this lorry after that. The driver of his lorry got out of his cab and went over to him. He was just going to give him a length of his tongue, I think, when he saw what sort of collar he had on he ended with saying, "You ruddy nearly met your Governor that time, didn't you.

George himself, it must be admitted, enjoyed a well-deserved reputation as a less than skilled driver of the various vehicles he used for attending to his numerous widely scattered clients. When, in his later years he made a thoughtless manoeuvre which caused an accident, he was advised by the sensible local bobby that provided he promised never to drive again no charges would be pressed. 'Thank God for that!' was George's response and it came as a considerable relief to his family who were anxious for his safety.

Still another tale reached me from Kirdford and I believe it is true. Two chaps in Kent fell out over a piece of land, I think it was, and they couldn't settle their differences between them so they decided to go to court about it. One chap did have a solicitor and the other one didn't . This one chap went to the solicitor and poured out all his tale and his solicitor said. "I can't take this case for you but I'll recommend you to go to another one in the village and if you agree to go to him I'll write you out a letter of introduction." Which he did. The old chap pocketed it and off he went. He went right by the solicitor's place before he realised it, and he thought "I'm not going back there today. I'll call tomorrow or the next day." He looked at the letter and he thought "I wonder what he's put in it?" It gradually grew so that he got one of his sons to read it to him. And all that was in it was "Two geese are coming to town, you pluck one and I'll pluck the other".

Several summers ago I'd been looking round my bees and I found

I had a nice lot of sections on some hives at Kirdford. My sister-in-law was staying with us for a few days and I was making up some sections, folding the wood up and putting the wax in and she saw me doing this and putting it into a box. I showed her how it was done. I thought when I took the other sections off I may as well put some more empty ones on. So after the midday meal I went off with the newly made sections and I wasn't too long gone when I was back again with some beauties, really full honeycombs. One they hadn't done just properly, but there was a nice lot of honey in it and I said, "We'll have that one for tea." My sister-in-law looked at these sections and looked at them again. As many bee-keepers know you can put sections in the hive and a lot of bees. and perhaps they won't touch them that year or the next. Then perhaps the third year when they've made them rather travel-stained you do get them filled with honey. Of course we do sometimes come off better than that but that is often what happens to every bee-keeper that's been going. My sister-in-law looked at them for quite a time as I've said and then asked, "Did they put that in while you waited?" and she's been in the country all this time, and should really have known better than that.

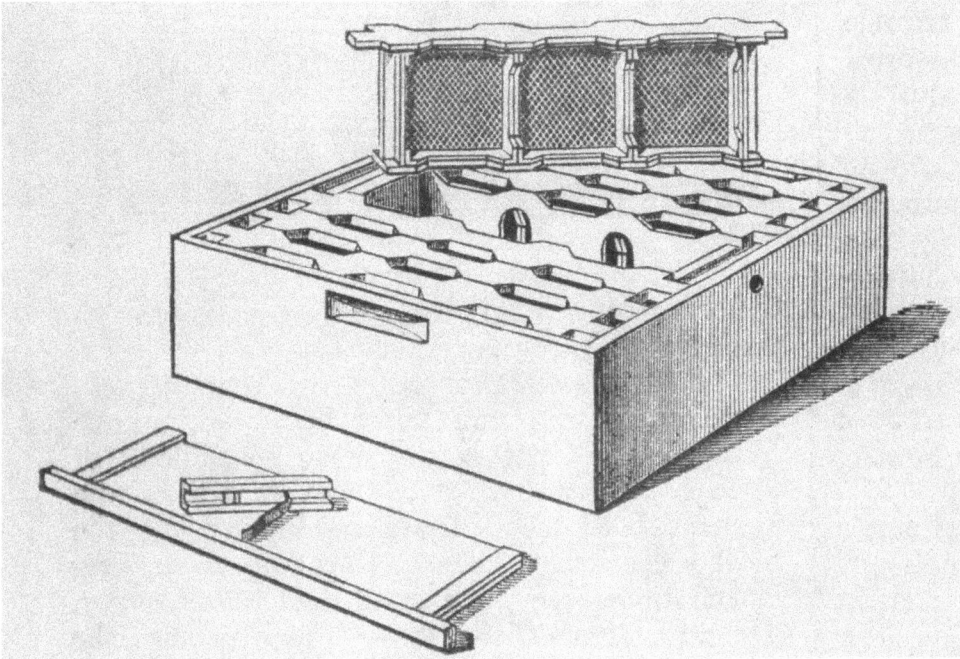

A section rack to make boxes of honeycomb

The fashion for getting the bees to build honeycomb in four and a half inch square basswood boxes, known as sections, for eating with a spoon and bread and butter, was popularized in Britain towards the end of the 19th century as a response to competition from the USA where bee-farmers exported them in great numbers. Though skilled beekeepers did manage to produce splendid examples in good seasons, the unreliability of the British climate and the strength of colonies needed to produce good clean specimens led to frequent disappointments. Beekeepers today have largely abandoned the struggle in favour of extracted honey, though lovers of natural honeycomb can still enjoy chunks of 'cut comb' sliced from large frames and packed into plastic containers.

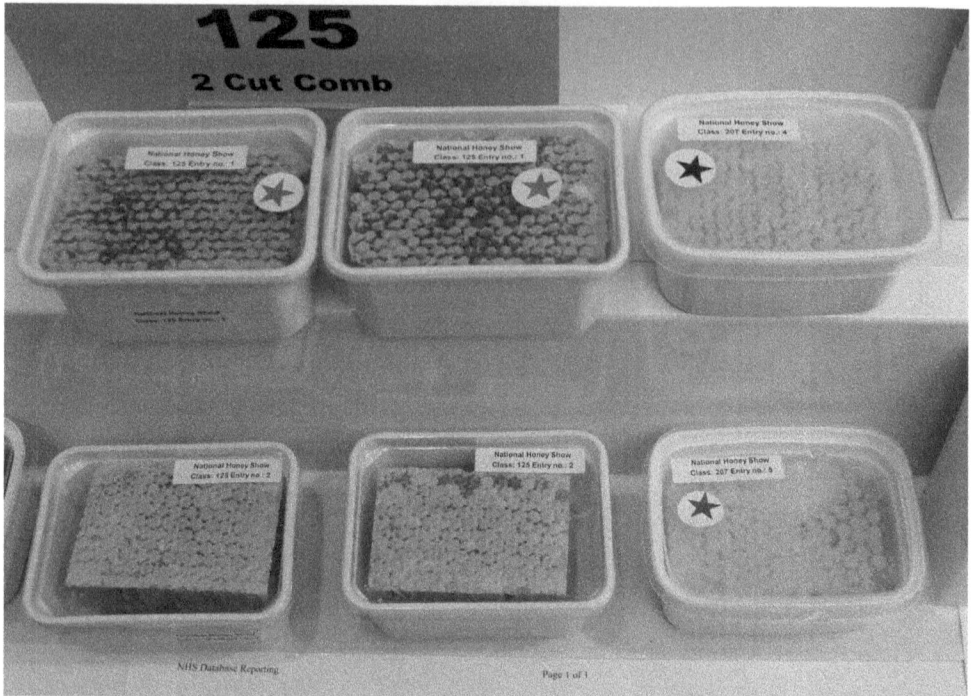

Cut-comb boxes - prize-winning examples

Here's another story about the Reverend Newberry, who had asked me to look after his bees for him. I took his honey for him one day just before the National Honey Show at Caxton Hall, London, and I said to him, "Don't come tomorrow for your honey because we shall be away in London all day, but you can come the following day," I said. "But there's one thing about your old book that I have my doubts about rather." "Oh!" he said "whatever is that?" He was rather concerned. And he said, "Well it's about old Samson and the bees. He had a lion come for him when he went to visit his Philistine girl friend and he took hold of its jaws and stretched its jaws out and killed it and threw it in the hedge. When he came by again in a fortnight's time he broke a piece of honeycomb out of this lion's mouth or whatever and took it along to his Philistine girl. Now a stinking old lion is the last place I'd reckon a swarm of bees would go in. Someone went along and saw them and thought it was bees and left it at that. But I reckon it was a lot of flies that had got there.

I don't reckon a swarm of bees would go into no stinking old lion."

Well, I happened to tell Mr. Wadey, the then Editor of Bee Craft, this by way of a joke, at Caxton Hall, and he looked up at me and said, "You've never been to Palestine, have you?" And I said, "No, I haven't." And he said, "Well, with the insects by night and the heat of the day that lion would be cleaned clear out in a couple of days. Well, actually, if that's the case where is there a better place for a swarm of bees to go in than the rib cavity under the waterproof skin?" And so it actually turned out to be that because this Reverend Newberry was very concerned about the matter and went home and consulted a large dictionary of the Bible which he had. He told me afterwards pretty well word for word what Mr. Wadey had said in Caxton Hall in London. One must give the old chap, the Reverend Newberry, his due for looking that out like that.

Herbert Wadey, known as Jim, was a celebrated Kentish bee-master from Crowborough. He was a year older than George and wrote several instructive and humorous books on bees and beekeepers. He was the most famous Editor of Bee Craft, the journal of the British Beekeeping Association.

An explanation of the story had been offered in the British Bee Journal by Mr. William Carr nearly a hundred years earlier:

Judges, Chapter 14. "And behold, there was a swarm of bees in the carcase of a lion". Samson had slain the lion. It is well known in those countries...that heat will in the course of twenty four hours so dry up the moisture of dead camels... that their bodies will long remain like mummies, unaltered and entirely free from offensive odour. Thus the lion's body formed a good hive for a swarm of bees and there were also the ribs of the lion for the bees to build their combs upon, and I have no doubt they built them straight (although they were not waxed) as Samson so easily "took thereof in his hands and went on eating". The 'ribs of the lion' was the first account we have of a bar-frame.

This myth has Greek roots. Aristeus was a minor god in Greek mythology reckoned capable of taming bees and keeping them in hives. He learnt from Arethusa how to sacrifice and batter cattle, then to encase their carcasses in clay in order to create new swarms of bees. This was an ancient Egyptian and Mediterranean practice, known as the 'bugonia process'. It is not recommended! Aristeus was credited by the Ancients with inventing a linen-gauze mask and the hive-smoking technique.

I was doing bees for the Wisborough Green Vicar, Vicar Williams. He was a Welshman and a great talker. One day I went down to his bees and found they were building Queen cells which meant I had to look all through the hive, find the Queen, clip her wings and cut out the Queen cells. Then I had to give her more room in the way of an extra super so that they wouldn't swarm again, or at least he hoped they wouldn't swarm again. That won't always stop them, though it will sometimes. Well, when I found the Queen cells were being started by the bees, I put a sack down on the ground and knelt on it, because I used to get the back-ache quite a bit those days. I was getting on with my job with my jacket off and my sleeves rolled up and all of a sudden I found the old Vicar was watching me from a distance and the only words he could say were "you amaze me, you amaze me." "That's all right, Vicar, as long as I don't make a mistake. If I make a mistake they're going to make me pay for it. Any forgiving to be done," I says, "you've got to do that." He went off with a bit of a chuckle and that was it. I just carried on with my job. Was his amazement because of the way I was handling his bees or because this was the first time he had caught me on my knees?

[Randall Williams, a former public schoolmaster, was Vicar from 1947 to 1981.]

I now have a story about the tenth lot of bees. This was in the British Bee Journal. It was in those days, a good many days ago, when the Vicar always claimed the tenth part. This old bee-keeper lived close beside the road in the village and he was a straw skeppist, as far back as that it was, and there were never more than nine lots in his garden. The old Vicar kept watch for a long time, but there were never more than nine lots. But one day the old chap had a tenth swarm and they flew all across the village so everyone knew about it. The parson said to his wife, "If I don't go and see him, you know,

I shan't get those bees." So they decided they would go and see him, and the old chap he said, "Yes, I've got a nice swarm of bees. I'll be along with them tonight." And when it got dusk the old chap went along and picked his hive of bees up, for when it gets dusk you can carry a hive of bees without shutting them in if you are very gentle and they'll just hang in a cluster inside. Then he went and knocked at the door and the Vicar came and opened it. Almost immediately he shot the bees out of the skep into the passage along with the old Vicar and he said, "The bees be thine, but the skep is mine." He made his tracks away. I don't just know how the old Vicar got on with them.

This story is a variant of an old yarn in which revenge is taken for some offence or claim on the beekeeper which he deems unjustified. George had many clients who were men of the cloth. He mentions a half dozen by name and his own story is laced with words and references to Biblical texts which echo the main literature which would have come his way. In Victorian times Church of England clerics and Non-conformists alike were in the forefront of developing beekeeping practice, motivated by philanthropic hopes of helping their poor parishioners as much as personal profit and satisfaction. Rev. Cyril Brereton of Billingshurst and Sutton for example ran a business selling queens and swarms. Scores of books for children with moral Christian messages based on the busy bee were published. It is evident from George's many dealing with the local priests that this tradition persisted throughout the 20th century.

Now here's a good tale from North Chapel which is actually true. I played cricket against a chap called Benny a good many times and there was no doubt he could hit a cricket ball too, but this day he was leaning up against the wall of The Swan waiting for it to open, and while he was waiting there with some others the Salvation Army came and set up just across the road playing and singing. Some of the Salvation Army people came round and started talking to these chaps looking on and one of them started to talk to Benny and said "Come and join us and have a sing-song along with us." Benny said, "No" he didn't think he would. The chap still persisted with him and he asked if there was something he regretted having done or not done. Benny said there was one thing. The other chap said, "What was that?" - he'd thought he'd got a little bit of a foothold, I suppose. "Well, I'm very sorry to think that I didn't drink more beer when it was twopence a pint."

An old friend of ours, Mr. K. Leggett, was ill in bed. He was quite an old character and he wanted to know what the doctor had said about him. He managed to creep out of bed and get his ear round the corner of the bedroom door over the stairs. When the doctor got downstairs his wife said "Well, what do you think of him, Doctor?" "He's very ill, he's very ill indeed. You'll have to watch him tonight pretty much," he said. "I'm afraid he'll go off in his sleep." And the old chap said he took jolly good care he never went to sleep that night!

Here are some tales from Petworth. An ordinary working chap there called Knight, who drove lorries for Sadlers, used to have some quaint sayings and doings. One day it was time for him to go off to work in the morning and he couldn't find his scarf anywhere. He got his wife looking for it as well and for some time she couldn't find it. Then finally she looked up and she said. "Well, you've got the thing on." "Oh!" he said, that's a good job you've found it, 'cos I'm blowed if I shouldn't have gone without it." Then he had a chimney fire one day. I suppose they got the Fire Brigade, but he got on his back on the floor and started edging himself towards the fire.

And his wife said to him, "Don't you look up that chimney like that. You'll get the fire in your eyes." "Oh," he said to her, "you don't think I'm going to look up the chimney with my eyes open do you?"

Now there's one about a doctor and a carpenter. This was one told me many years ago. The doctor was watching the carpenter making a window-frame and he'd got it made and was fitting it together nicely. He'd cut one piece just a little short, and tap as he would it wouldn't go up together. There was a little gap there. So the old carpenter he got his putty out, jammed it there as nice as anything and put a little drop of paint over, and you couldn't see a thing about it. The doctor says to him, "It strikes me putty and paint cover up a good many of your mistakes." The carpenter looked up at him, "Yes," he said, "and I expect a pick and shovel cover ever so many of yours."

Here's one I heard just lately. I am not sure if it is strictly true or not, but it is a good story. A chap had crushed his finger in a car door, just the tip of his finger. He thought he'd better go to hospital with it, so he went to hospital and showed the Sister what he'd done. And she said, "Go into that room and strip your clothes off and get into a gown and come and sit on the form next to that man there." And he said, "But I've only crushed the tip of my finger." And she said, "Do what I tell you and don't waste my time." She was so commanding he went in there and did as he was told and came and sat down beside another chap there waiting. After a bit he started chatting to the man next to him. "This is a tidy how do you do." he said. "I've only crushed the top of my finger and I've got to come and strip off, sit down and wait here." The other chap said. "I don't think you have much to grumble about. I came here to put a washer on a tap!"

Now here's a story about another old chap I was telling about to Dr. Henderson when he had come to see me- nothing very bad with me, I've never had much very wrong with me, I don't think. Now this old man was over eighty and he'd never been to a doctor in his life and he suddenly had a bad leg come on, and do what he would he just couldn't cure it. So the doctor gave him a week's course of treatment, i.e. rubbing with liniment and so on. This didn't do any good, so the next week he changed the treatment. This did no good either and the old chap got bad-tempered. The doctor said, "Well, we must just blame old age." And his patient said, "Old age be damned. I've had this leg as long as I've had that one and there's nothing wrong with that leg."

When I was down at Lodge Hill a good many summers ago at a Bee-keeping course we had a Mr. Austin Hyde, who was the headmaster of a rather high-class school in Pickering in Yorkshire. He had already given us a good talk on bees and bee-keeping at one session and then later on he gave us another talk on the lighter side of bee-keeping. He told one tale about his own school when they were having a visit from an Inspector who was very, very strict, so much so that even the masters were afraid of him, let alone the children. On this day he went into one of the classrooms with one of the masters and he had a little chat with the master. Finally the master said "Here are the lads and you want to examine them and ask them a few questions, I expect. They are all ready for you." He stood out in front of the class and he looked at them and he talked rather loudly to them and he began, "They marched round and round the city until the walls came tumbling down. Who did that? You there?" pointing to one of the boys, and this little old boy stood up; he was nearly frightened to death and he didn't know what to say. Finally, he said, "Please sir, it wasn't I." The Inspector was so aghast that he looked round at the master as if to say, "Well, what do you think of that for an answer?" The master perhaps didn't understand it quite right and he stuck up for the boy. He said, "Well, he comes of a very good family. He's a very good boy here, and if he says he didn't do it, I don't think he did do it."

With that the Inspector went to see another master higher up in command of the school and he didn't get it right either and said the same as the previous master. Then he went to the headmaster and told him about it. And he said, "Well, we won't make a fuss about it, we'll have the blessed walls put up again."

Finally, here is one about a man of the name of Slater whom I knew in the Home Guard. One night we were on parade and 'twas a terrible wet night. There was an instruction course indoors in the hall at Petworth, when they were given all the Home Guard instructions as to what they should do and what they shouldn't do if the enemy came. One of the instructors turned to Slater and said : "Well, say, the enemy is down at Coultershaw Mill and they're coming this way. What steps would you take?"

"Oh ho," said Slater. "Ruddy big ones."

Jokes about the bumbling incompetence of the Local Defence Volunteers with their broomsticks for rifle drill and 12 bore shotguns were legion, even before they were renamed the Home Guard, and long before the TV sitcom *'Dad's Army'* preserved such yarns for posterity.

George had a special fondness for practical jokes. One of his favourites was to ask a gullible victim if they had ever seen a white bee. When they said no, he would take out a matchbox, giving it to them to hold. 'Open it very gently,' he would say, 'we don't want to lose her.' As the box was opened he would keep on emphasizing caution. Eventually the hapless subject would push out the tray, and surprise, surprise; there they saw in the bottom a large white letter "B". Even knowledgeable beekeepers could be caught out as they might think that George had a rare albino bee with white eyes or just a worker that had been foraging among the white pollen of water balsam.

One day he met the new policeman for Wisborough Green for the first time standing by the crossroads at Hawkhurst Court. He pulled up alongside him, saying how pleased he was to meet him. After a short conversation he pulled a matchbox from his trouser pocket, showing him a queen bee in it. He explained that he was going to a queenless hive and he was keeping her out of the cold. But with that the queen suddenly took off and vanished. The P.C. laughed, saying, 'That's the end of her then!' George just smiled and carried on the chat. Minutes later she returned, went back into the matchbox and settled down. 'Fancy her doing that!' said the P.C. George smiled again and said, 'Well, but she had nowhere else to go.'

The best loved jests are rarely original. Old favorites can be told and retold by a favourite raconteur and unfailingly raise a guffaw from an appreciative English audience. How often can the eccentric Eric Morecambe, the 'simple amateur', be heard besting some acknowledged professional as when Andre Preview is assured that he, Eric, is playing the right notes, but not necessarily in the right order, or that Ronnie Barker wants fork handles, not four candles. In a social context of largely unlettered country folk, candlelit evenings, from time immemorial, were enlivened by the fiddle, ballads, and not least by the reminiscences and story-telling of bards and tellers of tales. George and his yarns match the pattern. The man himself would not appear out of place as one of the striking personalities pictured by Geoffrey Chaucer as Canterbury pilgrims, the structure of which poem was based on the plan that each should tell a tale, earthy or enlightening, appropriate to the nature of the teller.

Jeremy Paxman, in encapsulating 'Englishness' in his Portrait of a People sums up that attitude of mind that made the English culture what it is – 'individualism, pragmatism, love of words, above all, that glorious fundamental cussedness'. How perfectly George's life exemplifies this 'idea in the mind' of the eccentric individuality of Englishness! He symbolized in flesh and blood the archetype of peculiar rustic genius often portrayed in fiction such as Hardy's reddleman, Diggory Venn, Dickens' Joe Gargery and Sam Weller, George Eliot's Silas Marner, and Tickner Edwardes' Beemaster of Warrilow. Such real life celebrities as dog trainers, horse handlers, sheepshearers, gardeners, steeple jacks and fell walkers, invested with some inner magic still constantly prove an astonishment and delight to bystanders. The control of bees, by its very oddness, its mystery and its inherent threat to life and limb, is a potent stimulant to such admiration even for quite novice beekeepers. When it is allied to an adventurous and resolutely abnormal lifestyle and an awesome disregard for the dangers of swarms of belligerent insects, as personified by George Wakeford, it cannot fail to captivate an audience. We British are still properly characterized as green-minded 'good lifers', wildlife and country lovers at heart and warm to one who gloats over the wonder of the scent of the first primrose and glories in an English garden. George's life, as Shakespeare put it:

'Finds tongues in trees, books in the running brooks,
Sermons in stones, and good in everything.'

George working a WBC hive

George smoking a hive

Authors of plays and novels generally provide their audiences with an appropriate background 'milieu' as the stage upon which their invented characters play out their adventures. In the best of them, memorable places interact with the action, enhancing and magnifying the characteristics of the people. Elsinore and Blandings Castles spring to mind, Egdon Heath, Exmoor and Candleford, the forests of Arden and Sherwood. Artistically these places are more than mere geographical contexts but have an appropriateness and symbolic affinity with those who inhabit them.

George was flesh and blood, not fiction. Yet one could not imagine a more fitting setting for his real-life saga than the region about Ebernoe Common and Balls Cross in deepest North West Sussex. This densely wooded, remote neighbourhood, at the heart of a semi-feudal ancient estate, surely ranks as the epitome of antique Englishness. It is a place where one might assume that neither the Romans, nor the Saxons, the Normans, the Industrial Revolution nor even Hollywood had done anything to disturb the unchanging way of life of the denizens of these villages. There are to be found the most species of bats in the United Kingdom, and areas designated as of special scientific interest. There too, in myth and folk lore, are told yet wilder fantasies. One would not be surprised to learn of witches and warlocks, elves, bandits and sturdy rogues, of Rumpelstiltskin and Tom Thumb. It is the site of the Horn Fair held on 25th July (St. James's Day) since the 16th century, now politely reckoned to commemorate the abolition of slavery, but far more redolent of some mysterious pagan past. People say that Ebernoe was once the haunt of 'midgets' and little people, that dwarfs were bought and sold, and that the stature of the local folk is an echo of this circumstance. Dwarfs were always in demand as pets or for peep-shows. Dealers in little people were reported to have used drugs and herbs to 'miniaturize' children. Shakespeare speaks of 'hindering knotgrass' when describing the dwarf, petite Hermia. To adapt an apposite Disney film title, 'Honey, they shrank the kids!' Another view is that the fair is 'the survival of a semi-barbaric custom of the times when 'gods' were made and burnt on village greens in order to avert calamities…probably…the sacrifice was a human one'. The historian Peter Jerrome, however, questions whether the whole concept is anything more than a self-perpetuating rural myth. Nevertheless he wrote of: 'the general air of mystery that always shrouded this remote and…inaccessible fastness…this improbable enchanted place.'

Messrs. Pullen and Rowland from Scratchings Farm bearing the horned sheep

Visitors need not be dismayed. The local people were, and remain, gentle, community-minded, worthy farming folk, gardeners, gamekeepers, parsons, pensioners and the like with a fondness for village cricket, stoolball and pheasants, rather than werewolves or hobgoblins. Yet it is such a romantic area that one can understand how that strange powerful personality George Wakeford, that eccentric magic man, with an exotic lifestyle, would, all his life, yearn to return to the place of his birth. The location so befitted the extraordinary nature of its remarkable native son.

It may be coincidence, but in Ebernoe churchyard is a memorial of a couple called Poste. Readers familiar with the great 1932 comic novel, 'Cold Comfort Farm' will recall that the heroine, Flora, is frequently named, "Robert Poste's child". It is reported that the lady so commemorated was in fact George's cousin. The milieu of the novel is well established to be in this general area of Sussex and may well have been familiar to Stella Gibbons. How easily one could imagine Ebernoe as the source of the fictional village of Howling, and Palfrey Farm as the home of the Starkadders! The area might well have been the inspiration for a host of rural grotesques and eccentrics, a company that could easily have included George as one of its saner members.

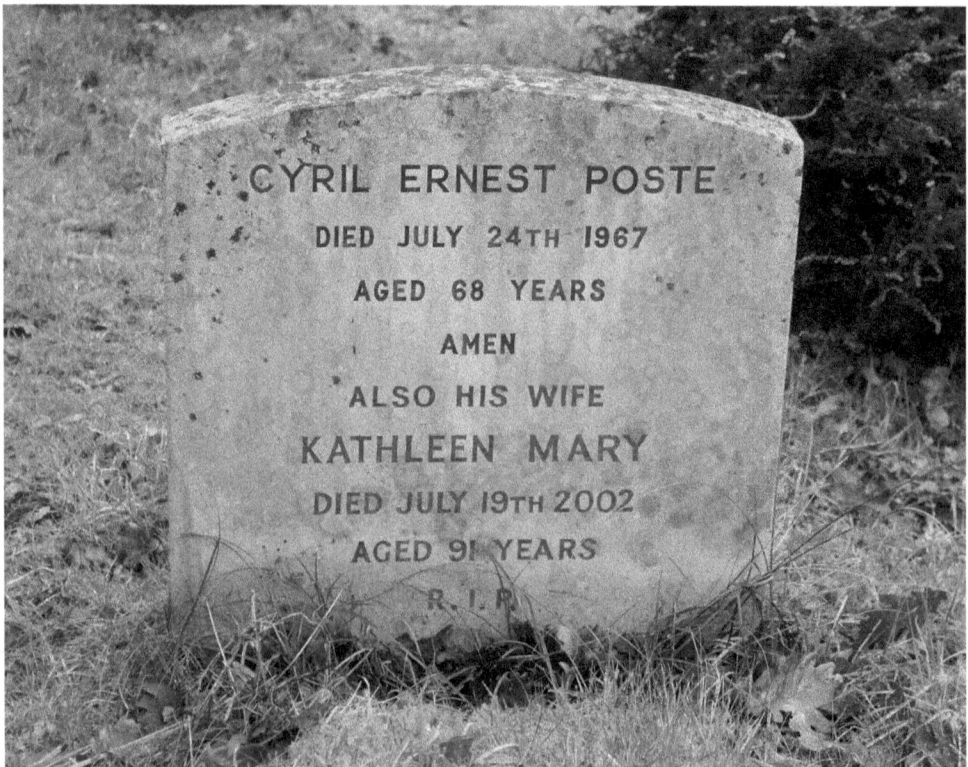

Poste gravestone in Ebernoe Churchyard

As for Wisborough Green, where George raised his family, it was then, and is now a shining example of an archetypal English village, with its cricket ground, church and old-world charm. St Peters has an interesting wall hanging which records the association with Canadian soldiery assembled for D Day in WW2 when Bren-gun carriers were parked on Ebernoe cricket field and pictures beehives to symbolise many parishioners' long standing relationship with that industrious insect.

Wisborough Green Church –Agatha Bowley

THE FINAL YEARS

George's final months were darkened for his friends and relations by the onset of Alzheimer's and Parkinson's disease. These afflictions reduced this admirably active and accomplished man to a sad and increasingly dependent remnant of the one-time wise and witty gentleman they loved and respected. He contacted pneumonia and finally gave up the struggle on 16th July, 1985, a date only a few days before Sussex beekeepers expect the main summer nectar flow to cease, save for the ragwort and the ivy.

His body was cremated at Chichester but his memory is still treasured by his swarm of friends and admirers.

Mr. Bill Hazelman wrote:

George was used extensively by the police and by the council to deal with swarming bees. I remember once we had a sunblind jutting out into the street at the Middle Street shop and a swarm had settled on the end. It was a public hazard; somebody might bump their head against it with serious consequences. George was sent for and in no time was calmly saying, "There she is" in that characteristic way he had. The queen was in a matchbox and George Wakeford had everything under control.

What most impressed me was his patience. When I was beginning beekeeping I'd go over to Wisborough Green with my many queries and he'd explain, and if necessary explain again and again. Other things too I remember, like going to the Honey Show with him last October and him bringing a box of apples to give away and from which trees he has struck many cuttings over the years. It was, of course, Benn's Red, that rare variety he had once been given. I remember too his knowledge of wild flowers and country lore....I remember the cap he always wore, the unbuttoned shirt and the way he called even quite senior people "boy". I always felt quite pleased to be addressed in this way.

There was a minute's silence at the Wisborough Green Association meeting in July. George Wakeford will be very much missed both for himself and for the help he so freely gave.

APPENDICES

Article written in 1961 for the Daily Mail "Get Ahead" Contest.

I - Bee-keeping in the Past

I have been a bee-keeper and a manager of bees since about 1917 when I had my first colony. My father and grandfather were also bee-keepers using the old straw skips as hives.

While they were successful bee-keepers in that they got honey nearly every year, their method of bee-keeping did not require the amount of bee knowledge that is needed with modern methods and I doubt if either of them would have known a Queen Bee if they saw one.

I read bee-keeping books throughout the first winter, and expected to manipulate and manage my bees the following spring and summer. But things did not go as I expected until our local postman gave me a demonstration on manipulation.

I soon picked this up and by attending meetings and further demonstrations arranged by the Sussex Bee-keeper's Association, I had now joined, I was able to manage my bees fairly well. Acarine disease killed many of my bees in the nineteen twenties, but by 1930 I had up to twenty colonies, when owing to a change in my work I could not manage so many and had to sell some of them. Since then I have gradually increased my stock and now have about 30 colonies. In addition to this I manage over 400 colonies for owners living within a 30-mile radius of Wisborough Green. When the petrol tax of 2s. 6d. per gallon came on I almost gave up trying to make it pay myself and my clients, but carried on, and, believe me, my margin of payment over their profit and my expenses was very small. Also I was over-worked and some bees were not managed as well as they should have been.

The Present

I was anxious and concerned about this and decided that improved methods were the only answer in my case. I have now invented a feeder that will cost only a few shillings and will feed the whole colony in one go. This will give them sufficient food in one feeder to last them a winter and they can take this in two days thus saving me many journeys in feeding time.

Again, wax extractors do not appear to have improved since my mother extracted wax previous to 1915. So I set about this and have invented a new type that has been sought by the editor of the British Bee Journal and Messrs. Robert Lee, bee appliances manufacturer, of Uxbridge, was interested. Patents for these two ideas have been applied for and a third is in the hands of an agent who will be given the word go when a complete model proves successful. In addition to this, should one or all of these three be successful I need a better method of clearing bees from their honey supers, a better method of uncapping this honey, also of extracting and straining.

George with blocks of wax, a valuable by-product of the hives

I also need a better method of swarm control and have one in mind to try out this coming season. Should this come off a book called Bee-keeping Simplified will not be needed and I hope a few sheets called Simple Bee-keeping will be far more helpful. Throughout all my bee-keeping years I have tried to be helpful to our little bees and their owners. Too few people are taking it up and bee-keeping is by no means in a flourishing state.

Forecast of the Future

Being almost forced to handle bees in weather that was not suitable for opening their hives or homes, has, I believe, given me a greater insight into their lives than few if any other bee-keeper and I feel that I can handle bees with any other person in England. I shall, however, be sixty-one years of age next birthday and I don't want to take what I know away with me. I would best of all like to make a film on bees. I brought this matter up at a West Sussex Delegate meeting at Worthing over two years ago only to be told that it would cost too much. I have seen a good many nice films on bees, but they all lack in detail that would help the beginner. For quite a while the Russians had the best film called "The Sunny Tribe" Then the Americans drew level with "Bees for Hire". Where has England got to?

I am sure I could plan a film to beat both of these, and one that all would wish to see. It would benefit bees, bee-keepers, crop pollination and thus general production of produce as well. This film may not cost the full £5,000 but anything left over would be most welcome for experimenting with bees in warm houses during winter and in many other ways.

NOTE. - This article was written fifteen years ago (1961) for the "Get Ahead Contest" in the Daily Mail. I am now teaching bee-keeping at the Weald School evening classes and I have enough film for two evenings showing. These need only the sound to be put on them, which I hope will be done soon.

Sadly, like other notable bee-masters such as T.W. Woodbury who introduced moveable frame hives to Britain and William Broughton Carr, George never published the book he was well capable of writing, and so was the less celebrated by his fellow enthusiasts nationwide. He was forever bubbling with ideas for the betterment of the craft, but lacked the necessary connections to get into print beyond this little autobiography. Neither did he have the means or up-to-date scientific expertise for the task. Equally his lack of any concern for making money or profit beyond what was necessary for his family needs ensured that his ventures came to little. He lost money for instance on his wax extractor which set him back £400 in agents fees for the patent, cost £4 each for Ron Flexman to design and Ernie Richardson to manufacture, and sold only a matter of two dozen for £4-10 shillings. As for his services, not only did he neglect to collect all his dues but also he charged for his time only two shillings and sixpence an hour (12 pence halfpenny). This seasonal work was thus no better rewarded than a basic agricultural worker, though he would have supplemented it from the sale of honey, wax, and equipment, and from fruit farmers for pollination hives and for occasional tutoring.

Given the right support he could have made that detailed instructional film for the benefit of posterity, but the opportunity was lost. His pre-eminence was most marked in his uncanny skill in actually handling bees and his mastery of the most difficult beekeeping manipulations in the most unpromising situations rather than astuteness at making money.

II-BEES WITHOUT STINGS

B hopeful, B cheerful, B happy, B kind,
B busy of body, B modest of Mind,
B earnest, B truthful. B firm and B fair,
Of all miss B haviour B sure to B ware,
B think, ere you stumble, of what may B fall,
B true to yourself, and B faithful to all,
B brave, to B ware of the sins that B set.
B just and B generous, B honest B wise.
B mindful of time and B certain it flies,
B prudent, B liberal of order B fond,
Buy less than you need B fore B vying B yond,
B careful, but yet B the first to B stow,
B temperate, B steadfast, to anger B slow,
B thoughtful, B thankful, whatever you B tide,
B just and B joyful, B cleanly B side,
B pleasant, B patient, B gentle to alls
B best if you can, but B humble pitfall,
B prompt and B dutiful, still B polite,
B reverent, B quiet, B sure, and B right,
B calm, B retiring, B never led astray,
B tender, B loving, B good and B sign,
B loved thou shalt B and all else shall B thine.

By E. E. B.

A FAREWELL TO PENNYGATE

Beside the Amaryllis by the pool
Basking in late September sun,
I ponder on the years at Pennygate,
The peace, the merriment and fun-
Good years that number over five and twenty,
Rich treasure store of happiness in plenty
The drowsy honey bees
Hum lazily among the autumn flowers,
And from the garden hive
We garner honey stored through many hours
From early crocus in the Sussex spring,
From apple and from cherry blossom trees,
From summer roses and from purple ling,
From honeysuckle and from buddleia.

All this wealth and sweetness from the years
Spent merrily at well-loved Pennygate
Are safely stored in my deep memories
And carried forward to my new estate.

Agatha H. Bowley

[Pennygate is a late Tudor house at Kirdford]

III·FLOWERS FAVOURED BY THE BEES

February	Aconite
	Crocus
	Hazel Catkins
March	Pussy Willow
	Celandine
	Dandelion
	Primrose
April	Wallflower
	Cherry Blossom
	Bluebell
May	Flowering Currant
	Raspberry
	Holly Flower
June	Roses
	Blackberry
	Sycamore
	Lime
July	Rape
	Willow Herb
	Knapweed
August	Thistle
	Clover
	Thyme
	Heather
September	Ivy

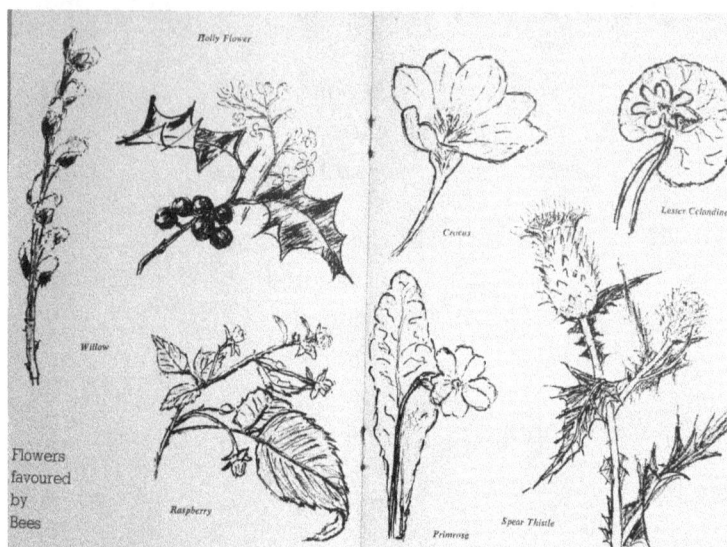

Flowers favoured by bees] (A. Bowley's sketches)

George does mention oilseed rape which was just becoming popular with farmers and was to prove a plentiful source of honey and pollen for the bees but a headache for the beekeeper. Rape honey sets rock hard in the combs so that even the bees cannot use it or the beekeepers extract it and it lacks colour and distinctive flavour. George wrote of it, 'Oilseed rape can be a good crop but the honey needs taking off very quickly as it crystallizes rapidly and loses its texture. I did have some however that was an admixture of wild flowers and that was excellent.' In commercial production the brilliant electric yellow fields are at their best in April and May, some weeks earlier than George states. At the time he wrote national production was only .06 million metric tons but by 1985 it had risen to 1.2 million. Mrs. Duncton recalls that in the early 80s George, still looking after hives at Palfrey, was pleased to put his bees in the experimental crop being trialed there.

 He does not list charlock. In his youth this wild mustard, quite similar to rape, was known as the 'glory of the cornfields', but George lived to see it banished as a splendid source of nectar by the use of herbicides on wheat, oats and barley. White clover which he includes had also become less dominant as forage for the bees because of the move from grazing to arable crops. He does not mention ragwort, deemed a pernicious weed and said to produce a bitter honey. It is, however, like ivy, a rich source of late nectar and pollen. The bramble or blackberry remains one of the most reliable local foraging crops. There is no certain record of George taking his bees 'to the heather' in pursuit of a late crop of honey from ling such as that found in the New Forest, and of course the West and North of Britain.

He did once collaborate with Mr. McWhirter, a local commercial beekeeper, who took a consignment of hives to the borders of Yorkshire and Derbyshire. Mr. McWhirter pressed out the thixotropic honey from his combs but George used a 'Perforextractor'. This is a bed of sharp needles fitted to a handle used to puncture the midrib of a comb and so agitate the jelly honey. It then becomes liquefied enough for most of it to be spun out in a tangential extractor. His emptied combs could then be reused.

Acknowledgement of References

Petworth Society Bulletins. No. 31 -1983, No. 41 -1985

Under the Greenwood Tree Thomas Hardy

The Men with Laughter in their Hearts Peter Jerrome and Jonathan Newdick

Not all sunshine hear -A history of Ebernoe Peter Jerrome

Ebernoe Church of England School - A History Frances Abraham

Snaps of Parish History. Kirdford Parish Magazine 1970 – G.H.K.

Billingshurst – Linda Lines

Bee-master George Wakeford 1977 and Agatha Bowley. Regency Press

Encyclopedia of Beekeeping Morse and Hooper

Alphabetical Guide for Beekeepers Vol II Ken Stevens

Cheerio Frank, Cheerio Everybody Frank Hennig (F. Streeter)

The English Jeremy Paxman

Cold Comfort Farm Stella Gibbons

The British Bee Journal - 1873 onwards

Google search, Wikipedia etc.

Rural Life in Victorian England . G.E.Mingay

Taylors Centenary Catalogue 1880-1980

George Garland's 1930s photographs

Agatha Bowley who marshalled and edited the words of George's life story as he compiled it in 1976 and provided the original sketches.
Valuable advice from Jeremy Burbidge, Roger Patterson, Ann and Rodney Tyrrell, Roy and Josie Curtis [and the loan of photographs] John Griffin, Helen Abbott, Janet and Chris Duncton, Derek Adams and the patient technical assistance of my wife and daughters.

A note about the editor of this revised version of the Wakeford story

Author pictured

Geoff Lawes was born in 1930 at Castle Acre, West Norfolk, the son of a farm manager. He spent his schooldays in Suffolk and East Sussex. After wartime schooling at four different Grammar Schools, mainly Sudbury, National Service and a spell as a junior clerk for the Royal Norfolk Show and as a farm hand, he trained as a teacher at Goldsmiths College and subsequently Birkbeck College. Duly qualified he taught in Catford and Sedgehill Schools and spent a year in Fargo, -North Dakota as an Exchange Teacher. In 1968 he became Headteacher of Roger Manwood School, Brockley and five years later returned to his rural roots as Head of The Weald School, a mixed Comprehensive for 1500 students. He retired in 1991.

He was drawn into beekeeping by Allan Dugdale, Head of rural studies and teacher-mentor of Roger Patterson who collaborated enthusiastically in the preparation of this book. Beekeeping was for many years a lively club activity for many youngsters, with several prizes achieved at the National Honey Show. Geoff built an 1100 volume collection of books on bees and beekeeping and, encouraged by Jerry Burbidge of Northern Bee Books (NBB), prepared '*The Bee Book Book*'' a manual offering advice on the making and maintenance of a specialist library. He later revised and illustrated Irmgard Diemer's German text *'Bees and Beekeeping'* and wrote a comprehensive catalogue of his own collection, with notes on the authors and their written legacies to the craft prior to its sale in 2010. Meantime he owned or supervised up to 40 colonies, undertook lecturing on bees, conducted classes in hive-building, acted as Photographic Judge for the National Honey Show and prepared numerous slide-tape instructional programmes for the BBKA loan library.

In retirement in 2005 he compiled the 50 year *'History of the Weald School'* which includes a trenchant critique of the blunderings and interference of politicians in attempting the cheap betterment of English education by manipulating organisational structures, curricula and threatening instruments of 'accountability' rather than by investment in teachers, premises, and teaching materials. In 2011 NBB published his *'Victorian Beekeeping Revolution'*, a distillation of the story of the modernisation of beekeeping practice from 1873 to 1901 as recorded in the pages of the 'British Bee Journal'. This volume offers pleasurable bedside entertainment for aficionados and lay readers alike.

Once active in the Wisborough Green BKA he is now a dormant member and has finally retired from local government as a County and District Councillor. He is now reduced to three colonies and a life of gentle gardening, non-competitive golf, a glass of wine, re-reading good books, the comfort of his converted cowshed home and the love and adventures of his wife Gilly and their four children and entertaining grandchildren.

www.ingramcontent.com/pod-product-compliance
Lightning Source LLC
Chambersburg PA
CBHW062028210326
41519CB00060B/7199